T0295197

Spiritualizing the City

Urban spaces have always functioned as cradles and laboratories for religious movements and spiritualities. The urban forms a central and nourishing agent for the creation of new religious expressions, and continually negotiates new ways of being spiritual and establishing spiritual ideas and practices.

This book explores the intense and complex interplay between the (post) modern city and new religious and spiritual movement, bringing the city and its annexes into the foreground of current research into religion. It develops a new, ethnography-based analysis of the ways in which the pluralist experience of the "urban" inscribes itself into various religious practices and vice versa: how do religiosity and spirituality appropriate and transform meanings of the urban? It focuses on new religious expressions, cosmologies and ways of life that go beyond established belief systems and religious understandings, and explores new conceptions of the word "urban" in a world of increasingly extended urban environments. The book examines how cities are both considered as sites and sources of spirituality, where the globalization of religions takes place as well as the fact that globalization is linked closely to the process of localization. The socio-cultural and political uniqueness of the specific urban context are analyzed to present an innovative perspective on how the interplay between the urban, spiritual and religious should be understood.

This book brings a timely new perspective and will be of interest to academics and students in geography, sociology, urban studies, cultural studies and anthropology, as well as for urban planners and policy makers.

Victoria Hegner is Senior Researcher at the Department of Cultural Anthropology/ European Ethnology at the Georg-August-University, Göttingen, Germany.

Peter Jan Margry is Professor of European Ethnology at the University of Amsterdam, the Netherlands.

Routledge Studies in Urbanism and the City

This series offers a forum for original and innovative research that engages with key debates and concepts in the field. Titles within the series range from empirical investigations to theoretical engagements, offering international perspectives and multidisciplinary dialogues across the social sciences and humanities, from urban studies, planning, geography, geohumanities, sociology, politics, the arts, cultural studies, philosophy and literature.

For a full list of titles in this series, please visit www.routledge.com/series/RSUC.

Technologies for Sustainable Urban Design and Bioregionalist Regeneration
Dora Francese

Markets, Places, Cities
Kirsten Seale

Shrinking Cities
Understanding Urban Decline in the United States
Russell C. Weaver, Sharmistha Bagchi-Sen, Jason C. Knight and Amy E. Frazier

Mega-Urbanization in the Global South
Fast cities and new urban utopias of the postcolonial state
Edited by Ayona Datta and Abdul Shaban

Green Belts
Past; Present; Future?
John Sturzaker and Ian Mell

Spiritualizing the City
Agency and Resilience of the Urban and Urbanesque Habitat
Edited by Victoria Hegner and Peter Jan Margry

Spiritualizing the City

Agency and Resilience of the Urban and Urbanesque Habitat

Edited by Victoria Hegner and Peter Jan Margry

LONDON AND NEW YORK

First published 2017
by Routledge
2 Park Square, Milton Park, Abingdon, Oxon OX14 4RN

and by Routledge
711 Third Avenue, New York, NY 10017

Routledge is an imprint of the Taylor & Francis Group, an informa business

British Library Cataloguing in Publication Data
A catalogue record for this book is available from the British Library

Library of Congress Cataloging in Publication Data
Names: Hegner, Victoria, 1971- editor. | Margry, P. J. (Peter Jan) editor.
Title: Spiritualizing the city : agency and resilience of the urban and
urbanesque habitat / edited by Victoria Hegner and Peter Jan Margry.
Description: Abingdon, Oxon ; New York : Routledge, [2017] | Series:
Routledge studies in urbanism and the city | Includes bibliographical
references and index.
Identifiers: LCCN 2016029559| ISBN 9781138930728 (hbk) |
ISBN 9781315680279 (ebk)
Subjects: LCSH: Cities and towns—Religious aspects. | Cults.
Classification: LCC BL65.C57 S65 2017 | DDC 200.9173/2—dc23
LC record available at https://lccn.loc.gov/2016029559

ISBN: 978-1-138-93072-8 (hbk)
ISBN: 978-1-315-68027-9 (ebk)

Typeset in Times New Roman
by Swales & Willis Ltd, Exeter, Devon, UK

Contents

Illustrations

Figures

Table

Contributors

Synnøve Bendixsen is a post-doctoral fellow at the Department of Social Anthropology, University of Bergen, Norway. She completed her PhD at Humboldt University and École des Hautes Études en Sciences Sociales and her thesis was published as *The Religious Identity of Young Muslim Women in Berlin: An Ethnographic Study* (2013). Bendixsen is also the co-editor in chief of the *Nordic Journal of Migration Research*.

Tricia C. Bruce (PhD, University of California Santa Barbara) is an associate professor of Sociology at Maryville College, USA, author of *Faithful Revolution* (2011) and co-editor of *Polarization in the US Catholic Church* (2016). Her forthcoming book explores the use of personal parishes in response to cultural, ideological and ethnic diversity among US Catholics. She also co-leads *The American Parish Project* and has conducted applied research for the US Conference of Catholic Bishops.

Eva Dick is a post-doctoral research fellow at the German Development Institute (GDI) in Bonn, Germany. Prior to her current assignment, she worked as a researcher and lecturer in the Department International Planning Studies of the Faculty of Spatial Planning of TU Dortmund University, focusing on migration-related societal change and its impact on spatial development and modes of urban governance. Amongst other activities, together with Alexander-Kenneth Nagel she has directed a research project on the governance of religious diversity in the Ruhr Area, Germany, looking at the governance of religious diversity and interreligious activities in the two cities of Duisburg and Hamm.

Claire Dwyer is a reader in Social and Cultural Geography at University College London, UK, where she is co-director of the Migration Research Unit. She has research interests in the geographies of faith, migration and multiculturalism. Her current work focuses on design, material culture and popular creativity in suburban faith communities, drawing on case studies based in West London and Canada (see http://www.makingsuburbanfaith.org/). She is the author of *Transnational Spaces* (co-edited with Peter Jackson and Phillip Crang, 2004) and *New Geographies of Race and Racism* (co-edited with Caroline Bressey, 2008).

Victoria Hegner is Senior Researcher at the Department of Cultural Anthropology/European Ethnology at the Georg-August-University, Göttingen, Germany. Her research focuses on the interplay between urban culture and (new) religious practices and sensualities. Her forthcoming book (transcript) on Neopagan witches explores the negotiation of religious visibilities and sensualities in the city context. Her recent publications include "Hot, Strange, Völkisch, Cosmopolitan. Neopagan Witchcraft in Berlin`s changing urban context," in Kathryn Rountree (ed.) (2015). *Contemporary Pagan and Native Faith Movements in Europe: Colonial and Nationalist Impulses*. Oxford: Berghahn, 175–95.

Gertrud Hüwelmeier is an anthropologist and senior research fellow at the Humboldt-University, Berlin, Germany. Her research interests focus on transnationalism, religion, gender, postsocialism, urban anthropology and media. She has published numerous articles and books and is the editor of the volume *Traveling Spirits. Migrants, Markets and Mobilities*, co-edited with Kristine Krause (2010).

Peter Jan Margry is Professor of European Ethnology at the University of Amsterdam, the Netherlands. Previously he was director of the Department of Ethnology at the Meertens Institute, a research center of the Royal Netherlands Academy of Arts and Sciences in Amsterdam. Now, as a senior research fellow at the Meertens Institute, his focus is on contemporary religious cultures, alternative healing practices, cultural memory and cultural heritage. He has published various books, volumes and articles in these fields, among them the edited volumes *Shrines and Pilgrimage in the Modern World: New Itineraries into the Sacred* (2008) and *Grassroots Memorials: The Politics of Memorializing Traumatic Death*, co-edited with C. Sánchez-Carretero (2011).

Alexander-Kenneth Nagel is Professor of Religious Studies at the Georg-August-University, Göttingen, Germany. From 2009 to 2015, he was Professor for the Social Scientific Study of Religion at the Ruhr-University, Bochum, Germany. His research interests include the civic potentials of religious migrant communities, religious pluralization and institutionalized forms of interreligious encounter and the governance of religious diversity in modern immigration societies. A recent publication is "Religious Pluralization and Interfaith Activism in Germany," *Studies in Interreligious Dialogue*, 25(2): 199–221.

Anna Niedźwiedź is an assistant professor at the Department of Ethnology and Cultural Anthropology, Jagiellonian University, Kraków, Poland. Her research focuses on lived religion, materiality and spatiality of religions, pilgrimages and visual anthropology. She has published books on the Marian cult in Poland, for example, *The Image and the Figure: Our Lady of Częstochowa in Polish Culture and Popular Religion* (2010), and on lived Catholicism in Ghana, *Religia przeżywana. Katolicyzm i jego konteksty we współczesnej Ghanie* (2015) (*Lived Religion: Catholicism and Its Contexts in Contemporary Ghana*). She has taught at universities in the US and in Poland.

Sarah M. Pike is Professor of Comparative Religion and Director of the Humanities Center at California State University, Chico, CA, USA. Dr Pike is the author of *Earthly Bodies, Magical Selves: Contemporary Pagans and The Search for Community* (2001) and *New Age and Neopagan Religions in America* (2004) and has written numerous articles on contemporary Paganism, the New Age movement, the Burning Man festival, new religions in the media, ritual, environmentalism and youth culture. She recently completed a book manuscript on spirituality, youth culture, and radical environmental and animal rights activism entitled *For the Wild: Ritual and Commitment in Radical Environmental and Animal Rights Activism* (forthcoming 2017).

Raphaela von Weichs is an anthropologist with a PhD in African Studies from the University of Cologne, Germany. At present she holds the position of senior researcher at the University of Lausanne in the European Research Council project "ARTIVISM: Art and Activism. Creativity and Performance as Subversive Forms of Political Expression." She has written numerous articles, including "Sacred Music, Sacred Journeys: What Makes an Event Postcolonial?" *ThéoRème* 4: 1–11 (co-authored with Monica Salzbrunn, 2013).

Mariske Westendorp is an anthropologist specializing in the anthropology of (urban) religion. She is currently a PhD candidate in the Department of Anthropology, Macquarie University, Sydney, Australia. The focus of her current research is on the relationships between religious, political and socio-economic aspirations in urban Hong Kong. Other research interests include urban anthropology, and the role of religion in development work and projects.

Daniel Wojcik is a professor of English and Folklore Studies, and affiliate faculty in Religious Studies at the University of Oregon, USA. His areas of research include apocalyptic beliefs and millenarian movements, visionary and "outsider" artists, urban ethnology, alternative spiritualities, healing traditions and pilgrimage. He is the author of the books *The End of the World As We Know It: Faith, Fatalism, and Apocalypse in America* (1997), *Punk and Neo-Tribal Body Art* (1995) and *Outsider Art: Visionary Worlds and Trauma* (2016), and has published widely on the topics of apocalyptic worldviews, apparitions and visionary traditions, nuclear age mythologies, trauma and art, rituals of mourning and memorialization, popular eschatologies, subcultural movements and vernacular artistic expression.

Preface

Both of us, as editors living in major cities, have experienced how the presence of the spiritual and the religious in the urban environment has been changing over the past decades: creating, adapting and attracting various forms of spiritual and religious movements, the re-creation of traditional practices, and also, among city populations, a growing need for "meaning (-fullness) and fulfillment." Confronted with rapidly expanding urban landscapes all over the world, we regarded it important to provide better insights on how *in practice* the inhabitants and especially the people related to those movements actually interact and constitute spiritual or religious forms or domains within the urban. We also wanted to understand how, reciprocally, those lived religions and spiritualities affect the extended urban environment. We decided to raise this issue for exploration in an interdisciplinary international conference, which we named "Religion in Urban Spaces."[1] For that goal, the Georg-August-Universität of Göttingen generously provided us with an opportunity to organize that meeting in 2014.

The present volume is the result of the presentations and discussions from the Göttingen conference and the editorial brainstorming afterwards in the process of composing this book. To that end, the variety of scholarly disciplines involved in this project brought new views to the fore, and also discussions on how to align to the chosen, more anthropological conference perspective, which had a strong practice- and agency-oriented approach. Moreover, the publishing editor and the reviewers of Routledge equally played a positive role in the melioration of this project. All this intellectual and academic agency brought us to a distinct focus that we wanted to have reflected in the title of the book. The introductory chapter forms the synthesis and rationalization of that process.

We hope that this volume will fulfill its roles in contributing to future scholarly, governmental and political debates on the vital importance and function of religiosity and spirituality within the present-day urban and urbanesque habitat. This volume should be read as a statement advocating for more research on the ways that religion is a central aspect of an urban way of life.

Finally, we want to express our gratitude to the Ministry of Science and Culture of Lower Saxony and to Kurt Hacker for providing the additional means to make the conference possible. Our colleagues Nadine Wagener-Böck and Michaela Fenske enthusiastically supported us in realizing the Göttingen conference logistically. We also want to thank Jayeel Serrano Cornelio from the Ateneo de Manila University and Karin Klenke from Göttingen University for moderating some of the conference's sessions with great intellectual spirit. Further, we are indebted to various colleagues who were involved in the creation of this book: first and foremost, our dedicated contributors, who sustained our persistent way of editing, and also Regina Bendix of Georg-August-Universität of Göttingen, who, as always, had the right sense for bringing people together for our conference. We are equally grateful to Victoria Bishop Kendzia for the correction of English, sensitively oscillating between necessary alteration and the preservation of the contributors' particular writing styles. Furthermore, we would like to thank Monique Scheer for reviewing some of the contributions.

Victoria Hegner and Peter Jan Margry
Berlin and Amsterdam,
April 15, 2016

Note

1 The conference took place April 9–11, 2014. All but two of the contributors were selected out of the 22 presenters at the conference. Mariske Westendorp, working on the same theme, was not able to participate in the conference and sent us her contribution afterwards. Later, we also added Synnøve Bendixsen to our fine list of co-authors.

1 Introduction

Spiritualizing the urban and the urbanesque

Victoria Hegner and Peter Jan Margry

The expression "Sunday assembly" will for most people at first instance refer to a religious gathering of a traditional and typically Christian kind. Since 2013, however, the naming of "Sunday Assembly" can be encountered in various countries and cities, but then couched with a different meaning.[1] It is the naming of a "group" that is displayed as a global and urban network of people, depicting themselves as "a global movement of wonder and good" and "a godless congregation that celebrates life." The Sunday Assembly's mission was initially described as "to help everyone find and fulfil their full potential," although it was again recently changed into a slightly less postmodern wording into "To help everyone live life as fully as possible" and "to wonder more" and "to help often."[2] The "Sunday Assembly" movement got off the ground quickly. While it has decelerated in recent years in terms of growth, it does continue to spread. Its members are wide-ranging in age: from their late twenties up to mid-sixties, mostly middle class in background. The movement envisions to create "godless" congregations in every town, city and village of the world where people want to have a congregation.

Reading their mission statement and vision, one might get confused as the choice of the organization's name will guide readers into the direction of religion, yet at the same the group emphatically states that it is a secular community, one without a god. Then, as mentioned above, the connotations of the applied wording are important, as their denial only concerns "God," which leaves the potentiality open in terms of religiosity and spirituality of any other taste than the one connected to (the Christian) God. Moreover, holding their gatherings on Sundays and calling themselves a "congregation," their mission resonates as a seemingly religious movement whether it is in Amsterdam, Berlin, or elsewhere. During one Sunday gathering that took place on a gray October afternoon in a community center in a residential building in Berlin, the ceremony appeared as an experiment to blend different religious and spiritual teachings and ritual elements together and fuse them with some modern skepticism and pop-cultural flair.[3] Approximately thirty people listened to edifying lectures (in lieu of, but not unlike, religious sermons) on the "relevance of virtues in life" and "the Dalai Lama's secular ethic." Attendees also joined in dancing to live music: popular songs by bands like the Police, Abba and Queen were performed live so that the participants could get into the "right spirit of community." Seeing how people rather timidly started to move according to the rhythms, one could notice how young, experimental and "unsettled" in ritual structures the Assemblies still are. It was obvious that it is still too new for some, and several of the attendees were here for the first time and it felt odd for them to freely dance around and get "loose" among strangers. Dances then changed, with (meditative) moments of contemplation accompanied by the sound of a singing bowl, and again with the chance to share personal stories on how everybody "is trying to give his or her best" to make the world a better place. The goal of these ceremonies, constantly oscillating between intellectual and bodily performances, general lectures and confessional tales, is, as it says on the website, for one to be "energised, vitalised, restored, repaired, refreshed, and recharged. No matter what the subject of the Assembly, it will solace worries, provoke kindness and inject a touch of transcendence into the everyday."[4]

Taking into account their practice of dealing with the fulfillment of life, we consider this to be a religious movement. Helping and finding one's "full potential" in a communal form, as the group emphatically propagates, is a mission that aligns other new spiritual movements, which resulted from a variety of "New Age" spiritualities and new religious movements that came into being in the past half century, often precisely dedicated to a broad quest for meaningfulness in modern life. Such cooperative movements are, however, paradoxically partly a consequence of the ongoing subjectification process within contemporary society, which has accelerated the deconstruction of the existing religious paradigms, changing them from a formal, traditional, or institutional kind into the present-day wide eclectic variety of personal believing (Saraiva et al. 2016).

As a contemporary example of spiritual seeking, the Sunday Assembly exemplifies quite well what this volume addresses: the urban and its peripheral life environment in the broadest sense, and the resilience of religion within the constructed human habitat that once was prophesied by scholars like Harvey Cox to become void of religion and gradually turn into a vast secular space. Yet as we all know, this has not come true at all, as an—initially—predicted seemingly secularization of society from its traditional religions has been superseded by the introduction of a great variety of religious forms and movements from other parts of the world and by new, eclectically created spiritualities, all serving in a partly global subjectification process of religiosity (cf. Heelas and Woodhead 2005). What is striking in the example of the Sunday Assemblies as that members see their mission not as a vanguard ideology for just city people, but that they express a deep felt desire for bringing their mission to as many countries as possible. They do that as much in the urban, the suburban and in a countryside that is "sensitized" for such kind of movements: the extended urban area of what we label later on in this introduction as the "urbanesque." The urbanesque is a space not consisting of a dense build urban cityscape, but of more rural-built habitats in the "hinterland" surroundings, but where modern and "urban" thinking or mindsets have supplanted more traditional rural lifestyles.

Religion as an urban way of life

The presence of religion in society and particularly the interpretation of the presence and practice of religion have been changing dramatically over the past century.

At the beginning of the twentieth century, ethnologists and anthropologists cherished the idea that "true" and "authentic" life and tradition and hence "true" religion were to be found in areas of relative isolation, deep in the remote outback, or in fishing villages on isolated coasts. This idea or approach was constructed in opposition to the notion that the city, especially since the start of industrialization and its effects on society, was perceived as a focal point of change and renewal and as a consequence as the cradle of secularization. With very few exceptions, the city was seen as the environment where apostasy and atheism could blossom, as a destroyer of the religiously bound social community, a community continuously distracted from religious practice by the temptations of the secular (McLeod 1996: xxi). As early as 1844, a clergyman in London—the world's largest city at that time and paradigmatic for the urbanization process—professed in one of his sermons "that the life of cities is essentially a worldly life" (Clapp 2012: 48, cited as well by McLeod 1996: xxi). As the proclaimed "religious crisis" seemed to intensify over the coming decades, particularly among the working class, as well as middle-class intellectuals, it was the American theologian Josiah Strong, who, in his widely acclaimed book *Our Country* (1885), was one of the first who designated the metropolis as the dangerous breeding ground for "evil" par excellence. A marked inclination towards mixed marriage was then regarded to be one of the main causes of the apostasy (Rogier 1945: 121). Strong saw the upcoming metropolis as a forceful development of irreligiosity, creating a threat to civilization at large. Almost a century later, religion indeed seemed to have somehow vanished from the cityscape and urban life altogether, as the American theologian Harvey Cox acknowledges in his highly regarded book, calling it programmatically *The Secular City* (Cox 2013 [1965]). There he professed that he could no longer see a role any more for religion in the urban environment. Although his argument was still much in line with Strong, he simultaneously differed in his perspective on the city in important and interesting ways. Along with various social scientists like Peter L. Berger in those days (Berger 1967; cf. Berger 1999), he put the idea forward that at this point in history, it was no longer Strong's urban temptations that formed the threat to religiosity, but that the cultural changes of the 1960s created a new plurality of movements, placing religious "truth" in perspective and hence producing cities that would become fully secularized, with formal religion deeply marginalized within the urban environment.

The resilience of the city as a spiritual place proved to be much greater, as urban spaces continued to function as innovative "laboratories" for new religious movements and spiritualities. This proved to be no different since the 1960s. Neither did religion disappear, as it only took on other forms and expressions (Luckmann 1967). The fact that nowadays cities and metropolises have become such vibrant centers of religious practice has led social scientists to gradually reject long-held assumptions about the secularizing effects of urbanization and to stress instead the strong agency and resilience of religion in the urban environment. Harvey Cox is again a good example within this context. As early as 1984, he confessed that he was mistaken about his earlier predictions and that formal religions surely were not marginalized in the world. This time he called his book

Religion in the Secular City and proclaimed that religion had returned to the public sphere of society. It had been cities, as Cox saw it now, which facilitated the forceful religious revival due to the city`s cultural liberality and its openness to social experimentation. Although Cox had been among several scholars at that time who questioned the main suppositions of the secularization theory, he was one of the very few who gave the urban locale such a prominent role for the future of religions. His book represents an important demarcation point within scholarly debate. As researchers started to acknowledge that religions seemed to be flourishing in society and on the governmental and political agenda, the topic of religion began slowly returning as an urban phenomenon on research agendas, after it had long been seen as diametrically opposed to city life and culture.

This particular change of perspective is, however, not solely due to changing perspectives on the process of secularization, but has to be simultaneously seen as an outcome of the success of cities and metropolises themselves as dwelling modes and the power of attraction for country dwellers and migrants. City or urban life has been more and more equated with modernity in a positive way and the city has become a metaphor for work and, moreover, a better and more fulfilled life. Some researchers label the process since the end of the 1960s, following a dramatic worldwide crisis of cities, as a (neoliberal) "urban renaissance" (Kaschuba 2015: 13; Punter 2010; Urban Task Force 1999; Porter and Shaw 2009). Before this crisis, cities and their downtowns were subjected to a fordist work and planning regime and subordinated to industrial production and modern traffic. They appeared as places for labor rather than desirable locales for living and cultural experiment. Inner-city districts were left to decay and were socially as well culturally abandoned by urban planners and politicians. It was movements like the "New Urbanism" which slowly brought a change of perspective in international Western urban politics, calling for the restoration of existing urban centers, and for varied, pedestrian-friendly neighborhoods and ecological practices (Jacobs 1992; Katz 1994).[5] It was this particular sense of diversity and community, as well as environmental activism reclaimed for cities, that has turned the latter gradually into sought-after neoliberal life worlds with their own character, charm and charisma (Kaschuba 2015: 13).

Overall, today, as an expression of modernity, more people worldwide live in towns than in the countryside. One can argue whether "urban" is then still a useful differentiating denominator as, for example, most of Europe's population (ca. 80 percent) lives in urban areas. Urban culture has become the mainstream way of habitation. This development has not only taken place in Europe, but also in various places elsewhere in the world; the urban and metropolitan has become the desirable standard in nearly every respect, hence constituting also for religious and spiritual renewal and experiment, a fascinating cauldron.

Particularly since the turn of the millennium—and again especially since 2010—we thus have been witnessing a near boom of new studies on religions as they are practiced and are multiplying in cities worldwide (Livezey 2000; metroZones 2011; Pinxten and Dikomitis 2012; Gómez and Van Herck 2013; Cimino et al. 2013; Day 2014; Becker et al. 2014). As scholars of religious practice and behavior as well as of urbanity, we nevertheless decided to bring the themes of the urban and religion together in a new way. First, we concentrate on religious innovations: on religious practices, worldviews, dialogues that are (still) rather experimental, often vaguely organized or only very recently institutionalized and still contested. We furthermore look at religious groups and communities that can be considered as a kind of grass-roots movement, created by people who are not necessarily officially designated religious experts and yet feature as powerful actors within the cityscape. We perceive these groups and innovations as distinctively urban. It is the city in its social, cultural and economic heterogeneity that grinds down traditions and makes way for alternatives; and it is again the city that provides a critical mass of people necessary to realize these alternatives.

In our understanding of the interdependence of religion and the urban context, we furthermore and explicitly merge perspectives of religious studies with a sociology/anthropology *of* rather than *in* the city (Hannerz 1980; Lindner 1993). Hence, this book is not so much about the variety of religious practices and innovations one finds in a city. Instead, the analytical lens focuses on the modalities of how urban environments in their socio-cultural distinctiveness and religion or religiosity are interacting with one another. In other words, we examine the processes of reciprocal exchange in the practice of religious movements and cityscapes in which they are couched. Our approach comes close to some ambitious projects over the last years, such as the "Global prayers" research group, which investigated "the renaissance of religion in the world's metropolises" over a period of four years (2010–14).[6] However, we expand the perspective on what can be considered as religious and begin to include more prominently modes and practices that are highly syncretistic, fluid and shaped by the "new age" movement (Hanegraaff 1996). In so doing, we start to transgress well-established and state-authorized or contested religious modes and theologies such as those of Christianity and Islam, two religious manifestations that constitute a central focus in studies on religions in cities, in addition to Buddhism and Hinduism. This is not to say that we abandon the perspective on Christian, Islamic and/or Buddhist groups and communities altogether. On the contrary, we take those groups into our analysis, since they remain important urban actors. However, we continually concentrate on ways practitioners go beyond and shift deeply ingrained cosmological and theological understandings and/or experiment with organizational structures and forms of public visibility. As we explain in detail below, urban as well as religious studies need to reflect more intensely on and broaden the concept of religions, so they are able to capture the multiplicity as well as discrepancy of religious practices that are so characteristic for cities.

Recent studies that emphasize and focus on the mutual agency of cities and religions in general concentrate on the ways religions become politicized and communicate respect to powerful, disputed and/or subversive players within the urban governance system (metrozones 2011; Becci et al. 2013; Beaumont and Baker 2011; Al Sayyad and Massoumi 2011). This approach is particularly present in the concept of the "postsecular city" mainly developed by the geographer Justin Beaumont and the theologian Christopher Baker (Beaumont and Baker 2011).

Critics have already pointed to its rather problematic "linear temporality": from the secular to a "postsecular city and age" (Lanz 2014: 24; Leezenberg 2010). Nonetheless, the concept offers a productive perspective on how religions, or—in a broader sense—faith-based organizations, create new alliances and bifurcations with the secular sectors in the city, often taking over social-political responsibilities, which urban municipalities are no longer able or willing to fulfill (Beaumont and Baker 2011: 1). Religions thus develop into indispensable actors of civil society, as much as they produce spaces for their ideas and values and thus shape the cityscape in significant ways.

We take this aspect of interplay between religion, the city and civil society into account. And yet we simultaneously shift the perspective significantly from the relation of the political and the religious in the urban realm towards the ways in which religions, religiosities and any forms of spirituality become embedded into the everyday routine of city dwellers. Hence, we look at religions in the city less as a political expression than as a specific urban way of life: as a manifestation of its transitory and fragmentary character, its specific sensuality, its fluidity and heterogeneous aesthetic, of the accessibility in diversity and diversity in accessibility (Hannerz 1980: 99). Certainly the idea of "diversity" and religious "plurality" gets politically instrumentalized within the urban context, as we show in this volume. They also work as the city's "cultural potential," to help, for example, in the social "integration" of migrants.

In this context, we sharpen our analytical focus on the city as a principal, multifunctional agent for change and renewal and draw attention to its socio-cultural and historical singularity as it models religious practices on site. As Robert Orsi has already pointed out more than a decade ago: "What people do religiously in cities is shaped by what kinds of cities they find themselves in, at what moment in the histories of those cities" (Orsi 1999: 46). Although the significance of local urban contexts in shaping forms of religions and religious expressions even in our global, highly transnational times, has long been acknowledged, only very few publications attempt to make this perspective explicit and fruitful in their analysis. Previous works mainly focus on religious groups as they engage with the politics of place-making and thus produce locality and furnish the city with a distinctive "religious topography" (Burchardt and Becci 2013: 12–17, esp. 17). Space, as a multidimensional construction, being physical, mental, social and likewise religious (Knott 2005: 127) is an important category for analyzing created forms of "locality," which is a major focus of this book. However, we extend our perspective beyond the category of space and also see "locality" produced and reflected through specific modes of organizational structures, as well as aesthetic expressions, behavioral codes and formulated norms and values (of ex- and inclusion). For example, as Synnøve Bendixsen shows in this volume, young Muslims in Berlin generate their collective practice of conservative Islam distinctively and in various ways as "multicultural"—including practitioners of various ethnic backgrounds. They thus forcefully localize their practice and create a feeling of belonging to the city—that is, of being a "local"—since for them Berlin's cultural specificity is per se lived "multiculturalism."

One of the reasons why the potency of local urban culture, history and social space are so seldom taken into conceptual and empirical consideration lies surely in the risk it bears in perceiving the city as a topographically fixed cultural entity. We see the city as a dynamic and heterogeneous socio-cultural, economic and geographic formation. We understand the city as a kind of nodal point or area, where, as the political scientist Michael Peter Smith pointedly wrote, "global and transnational networks of meanings, power, and social practice come together with more purely locally configured networks, practices, meanings and identities" (Smith 2002: 109). From the city, those networks spread out again. Territoriality of local culture is replaced by a kind of urban cultural fluidity, meaning an ongoing, yet distinctive change of relationships, webs of cooperation and roles. (Lindner 1996: 323; Hannerz 1980: 276).

Theoretically, we are inspired by concepts such as the "habitus of the city" (Lee 1997; Lindner 2003), "the cumulative texture of local urban culture" (Suttles 1984) and the "urban imaginary" (Strauss and Wohl 1958). All of them with some nuances emphasize the idea that

> . . . a city is not a neutral container, which can be arbitrarily filled, but a historically saturated culturally coded space already stuffed with meanings and mental images. It is these meanings and mental images which determine what is "thinkable" and "unthinkable," "appropriate" and "inappropriate," "possible" and "impossible." (Lindner 2006: 210)

Religious practices and expressions, as they globally and transnationally generate themselves, are part of this fluid yet specific urban culture, its meanings and (materialized) images. They reproduce them and simultaneously shift and remodel them. Sometimes, religious practitioners and organizations use specific cultural images of certain cities intentionally to create forms of (political) participation and visibility locally as well as globally. A very illustrative example is Anna Niedźwiedź's ethnography on the Catholic Church in Kraków in this volume. In various and highly creative ways, the Church authorities as well as Catholic "lay"-practitioners establish presence in the city and shape the cityscape by staging, remembering and re-inventing Kraków's cultural and historical image as the city of John Paul II: the "Pope's city." Creating new sacred grounds and symbolic spaces devoted and connected to John Paul II, they not only localize their practice, but define themselves and "their new devotions and spiritualities within a newly-opened global context." Discursively, they turn Kraków into an important Catholic city worldwide and in that way as well as into a successful touristic trademark.

Inquiring into the specific interplay between the cultural, social characteristics of urban contexts and religions certainly asks for a particular methodological repertoire. A useful approach constitutes a comparison between religious groups, practices, or initiatives (such as "interreligious dialogue") in two or more different cities.[7] We offer this specific and demanding research perspective in this volume through a contribution by Eva Dick and Alexander-Kenneth Nagel that concentrates on forms of "interreligious dialogues" in two different German cities. Similarities and differences do not simply show how embedded are religions in urban (political) culture and social structures, but also illuminate the diversity of practices and groups that are still often perceived as culturally cohesive. As our research experience further shows, when looking at religions in urban context, we should strengthen our—what Paul Stoller once called—sensuous scholarship (Stoller 1997). Religions in cities develop complex sensory epistemologies and we have to understand them in their urban specificity: how religions sound, smell and taste tells us something about the urban experience as well as constituting it. We will start to approach the role of the senses in the interplay between religion and the urban in this book, by looking particularly at the function of religious music and dance (in Part III of this book).

Generally, urban cultures as well as religions are such complex dynamic formations, deeply shaped by social, economic, geographic and historical structures that a study on their interplay principally demands the crossing of disciplinary boundaries in methodology and analysis (Lindner 1996: 323; Coleman 2013: 50). We do so by bringing together researchers of different academic fields in this book. Our disciplinary scope encompasses geography, religious studies, ethnology, urban planning and cultural anthropology. We are—as the modern city itself—quite heterogeneous.

Therefore, we often do not share the same conceptual and/or descriptive vocabulary. Our perspectives sometimes differ and then converge again in astonishing ways. The particular diversity in our academic backgrounds—challenging as it is—seems to us to be the central way to capture the interdependence of religions as they are practiced in the city in all its complexity.

In this context, the theoretical paradigms, laid out above, brought us—in our differences—together and raised basic research questions, which ground and traverse the volume's contributions. We find each other on the same course due to our general interest in how exactly the specific urban culture inscribes itself into new religious and spiritual views, performative acts and behaviors. And vice versa, we ask: do religious practices and new forms of religiosity recast the meaning of the urban space in any specific way? We concentrate our perspective on four aspects. First, we inquire into cities' ascribed cultural and religious *plurality*. We are interested in the ways it is managed, re-imagined and thus remodeled by religious practitioners and "experts" themselves and in how this *plurality* is being utilized as an important socio-cultural and political resource by city authorities. Second, we focus on the role of *migration* and the relationship between the urban context and transnational mobility. How are religious practices not only transformed, reinforced, or newly invented in the city and thus localized, but also, in which way do migrants generate themselves as global actors and preserve transnational ties and thus link the local with the global? Third, we look at the function and creation of *space*. How do religious groups negotiate (sacred) space and visibility in the urban context and what kind of urban imaginaries come into play? As we dive into the interplay between religion and the city further, last but not least we ask: can the physical *body* be an agent of creation of (sacred) places and spaces within the urban setting? What role do ritual movements, dress codes, dancing and visualizing emotions play within this context? In this way, we inquire into the function of material culture/materiality of the city including religious/sacred soundscapes as well as architecture.

In order to bring such queries from our theoretical paradigms to an answer, we first need to go into the main concepts used in this volume.

The spiritual, the urban and the urbanesque

The particular interdependence of urbanization, urban space, and religions and spiritualities has returned to the research agenda of scholars, and yet is still under-theorized. Therefore, one of the primary goals of this book is to develop a new threefold theoretical vocabulary for analyzing the relationship between the urban and religious beliefs and practices.

The first keyword to address the relationship is "spiritualizing," as it appears in the main title of the book. This single active verb points at the outset of the volume to our praxeological concept: spirituality and religion are something people *do* and constitute. Spiritualiz*ing* in this book refers to performative acting growing out of and taking place as a result of the experience of the urban environment. "Sacredness" and/or "sacred spaces" in the city are active constructions that take (visible or secretive) presence in the city.[8]

Furthermore, "spiritualizing" refers to our idea that, in contemporary religious processes, a distinction between religion (religiosity) and the spiritual is becoming gradually problematic. To that end, this book employs the twinned concepts of religion and religiosity as broad and deliberately blurred analytical concepts that cross the boundaries of traditional institutional religions. "Religion," as we understand it, also refers thus to new, alternative, or implicit forms and movements of religion and spirituality. We regard it better not to differentiate between the two, as in practice, people, especially within the Western urban habitats, often make no distinction between religion and spirituality themselves anymore. More and more, they even refer to themselves as in the popular adage "spiritual, but not religious" (cf. Fuller 2001). This stance implies that we do not take "religion" and a fortiori "religiosity" as the constrained form of the mainstream institutional religions, but as the inclusive wording for all forms of practiced religiosity of which also the spiritual, the new spiritual and New Age movements make part of, including implicit forms as the subjectified quest for meaningfulness in life (cf. Luckmann 1967; Bailey 1999; Partridge 2004; Carrette and King 2005; Knoblauch 2009; Mohrmann 2010; Margry 2012; MacKian 2012). From this inclusive and encompassing perspective, we were able to focus through a global lens at the wide variety of religious practices within the urbanized domain. Today's religious studies need this broad perspective and paradigm on religion in order to deal successfully with the nearly infinite amount of religious expressions and hybrid forms which can be encountered in cities and their outskirts.

The expressions dealt with in this volume thus range from traditional Catholicism as it is re-invented by devotees in the face of urban demographic changes, immigrant Asian and African religions, new—fundamentalizing as well liberalizing—trends in Islam, to New Age and implicit sensory-related forms like music and dance.

The second important keyword in this study deals with the city's "urban" character. This adjective seems at first instance rather clear and distinct. "Urban" is mostly equated with "(conflict-laden) diversity," "creativity," "innovation" and "cultural openness." We also use the term in this way. The city as its territorialized (material) expression, again, has often been described as a paradigmatic site of modernity. Those ascriptions are formulated in juxtaposition to the contrasting background of what is supposedly "provincial," "rural" and characteristic of a "village." However, in various parts of the world, the sharp distinction between cities and rural areas is vanishing, both culturally and topographically. Suburbs develop rapidly, metropolitan regions expand and rural spaces, as some researchers proclaim, are largely disappearing (Schmidt-Lauber 2010: 15; Kersting and Zimmermann 2015). When looking at and investigating the "urban character" of religions and spiritualities, we have to take those developments into account.

Hence, we must widen our category of the *urban*, and bring in a concept that is tightly linked with and formulated by the results of qualitative ethnographic research on location. Ethnographic research points out that present-day agglomerations and conurbations and their transition areas no longer match the traditionally inscribed idea of the urban as a phenomenon solely bound to densely built city centers and cityscapes. We experienced through our research that lively and innovative developments in the religious domain diffuse from city areas into wider cultural-geographical regions. Again, innovations often originate and blossom precisely in suburbs, in sprawled residential districts and in former industrial and business zones. It is from such sites that they pave the way for religious innovations in the city. This is partly due to a significant shift of the urban phenomenon of international migration and the ways incoming migrants settle in cities. The famous "ring" or zoning model of the Chicago School of Sociology (Burgess 1925)—according to which migrants, after their arrival, first find "home" in inner-city districts and then move out to the suburbs as they climb up the social ladder—is to some extent obsolete today. Because of high rents in now often gentrified inner-city areas as well as due to bureaucratic settlement politics of municipalities, migrants today often settle first in the outer ring of the city or even beyond (Hegner 2008: 51–77, esp. 54–6). It is then from there, from the outskirts, that they build their faith communities and start to create spaces for their religious practice, thus establishing pluralized topographies of faith. Furthermore, as Claire Dwyer illustrates in her contribution in this volume, migration has long reached the suburbs and sometimes international migrants even constitute the majority of their inhabitants. The suburbs and not the city, demographically speaking, work as gateways to create forms of impressive visibility of their religious beliefs. It is there where migrant faith communities build temples and monasteries "by activating transnational networks and flows of finance, social remittances, objects and people across national borders," as Dwyer writes. Thus they "offer new forms of sacred geographies in the city, and in particular a distinctive engagement with suburban landscapes."

Comparable processes take place on those edges of suburbia where one places transitional zones into the rural and the semi-rural. There one finds (new) towns and villages and residential districts not (yet) taken in the process of conurbational absorption. Socially speaking, in such areas people commute and have a strong orientation to the city. Although such areas are not physically and geographically strictly urban, nor do they have an urban appearance, but the *living culture* one finds there and the residents' mentality are urban oriented (Korff 1985; Lindner 1996: 325).

As academia is in need of a more complete understanding of what is nowadays actually happening in religious everyday life practice, we became inspired to broaden the analytic framework of the ongoing debate and to engage with a new concept: the "urbanesque." This is a new term which we use here alongside "urban" as a complementary domain in order to understand religion as it is expressed in urban places. On its own, the "urban" category can no longer account for the cultural-geographic dynamics of the city environment and its urban mentality and living culture. Such living cultures and mentalities supersede the traditional physical "urban landscape." The concept of the "urbanesque" widens the analytical scope for the intense *interaction* between the urban cityscape in the strict sense and the expanded—urbanesque—environment. It also allows one to conceptually frame the specificity and diversity of small and mid-sized town life as well as suburbia, exurbia and the "urban" rural periphery, this last category being areas which are inhabited by commuters or former city inhabitants, who after having moved out, actually continue to practice urban values and lifestyles or have taken up an urban lifestyle.

The proposed category of the "urbanesque," as we see it, hence, provides a more complete idea and representation of what urban means today, and of where urban life stretches and is practiced and, for our purposes, where (also) new developments on the religious take place. It thus brings zones of important religious innovations into focus, which are usually overlooked in urban studies, as much as in religious studies. With this concept in mind, one finds striking examples of how residential and industrial zones or suburban regions become the laboratories for spiritualizing the city and its urbanesque environs. It illuminates how new religions or movements entering a city are able to recast and spiritualize the urban environment. Let us succinctly demonstrate this with the two metropolises that the editors live in: Amsterdam and Berlin. Our empirical and analytical vignettes take very different points of departure. We aim less for a comparative perspective; instead, we want to show the diverse ways one is able to approach the phenomenon of "spiritualizing the urban and urbanesque."

For Amsterdam, we focus on an exemplary site: the modern residential Bijlmer or Amsterdam South-East district (85,000 inhabitants). It was built as a new neighborhood in a polder on the border of Amsterdam in the 1960s and 1970s, and reflects the urban planning point of view of those days, when large numbers of city dwellers were resettled outside of town in completely new residential areas, which mainly consisted of modern flats. Bijlmer reflected the idea of a semi-rural new town, with lots of greenery and a fully separate road system with dedicated parking isles. The planners' views were totally in line with Harvey Cox's ideas on secularization, and thus the vast new town was built without any church buildings. And as often happens with model towns, Bijlmer soon proved to become outdated with the first residents beginning to move away, leaving empty apartments and parking isles behind. From a modern "ideal" middle-class settlement, it turned into a lower-class ("black") migrant quarter. The district soon became a very multicultural one, with most of the inhabitants originating from the Caribbean area and Africa. Hence it became a most productive neighborhood for the residents' churches and various religious movements. It proved to be fertile soil for the accumulation of religious groups. Over the years, some 120 migrant church communities, of which 60 were more informal with a high mobility, moved around or found space in the Bijlmer district (Oomen and Palm 1994).[9] The low-budget movements consisting usually of car-less immigrants who squatted the empty parking constructions and used basements and storage rooms, creating a whole new network of makeshift churches. Schools, community houses, sports halls and private houses also supplied the needed sacred space. Dozens of new and African and Caribbean religions were often forced, due to cost and lack of space, to share the same religious building, very much like what Michael Owen Jones found in the Los Angeles agglomeration, that is, hundreds of "storefront" churches that dot its urban sprawl, and which are to be seen as a front line of religious activity. These indicate how urban areas are functional places designed for living in coexistence, but also providing refuges for those in need of the consolation and support of religion and the spiritual life.

The situation in the Bijlmer gave the municipality of Amsterdam the idea to build a dedicated church of a multifunctional and multiconfessional design. The building *De Kandelaar* ("the candle holder") houses 15 African denominations (Van der Valk 2014), which were called "Church Centers" instead of churches.[10]

This solution was arrived at partly due to financial limits, but it was also the result of a governmental policy regarding the constitutional freedom of religion and to bring religions in coexistence with one another, in order to come to know and respect each other. But reality proved more difficult, as the African churches involved were often more committed to their improvised buildings and not to formal issues of accountability and they wanted to return to their makeshift churches. In its slipstream not only did the suburban Bijlmer district become spiritualized, but also the older residential districts somewhat closer to the city center of Amsterdam.[11] Such cases show how the urban can unexpectedly facilitate and affect religious expressions, while reciprocally the religious movements affect the cityscape and create community and social cohesion.

Shifting the focus to Berlin, we leave the analytical unity of an exemplary site and take up Peter L. Berger's now famous assessment of Berlin as the "world capital of atheism" (Berger 2001: 195). Focusing solely on Christianity—as Berger does—one can certainly conclude that Berlin is rather irreligious and non-spiritual. Only 30 percent of its inhabitants are members of the nationally recognized two

Churches: Catholic and Protestant.[12] Their numbers are shrinking rapidly, as are the number of buildings—chapels, community buildings, churches—maintained by the two major religious organizations. In 2005, there were still 578 of them, today the number is below 400 (*Statistisches Jahrbuch Berlin* 2005: 164; *Statistisches Jahrbuch Berlin-Brandenburg* 2015: 168). Only 5 percent of the Christians in Berlin and its immediate surroundings (called Brandenburg) attend Mass (*Statistisches Jahrbuch Berlin-Brandenburg* 2015: 168).

The mostly empty churches are protected under the law of listed historic buildings, which means they must be preserved for future generations and cannot be torn down. Hence, they often are re-designated as functional rooms for receptions and galas, and marketed as "exclusive event locations." Conversely, they are reassigned as buildings for social relief organizations or as art galleries.

From this perspective then, the city and its urbanesque habitat are being heavily de-spiritualized. Yet, if we look beyond Christianity, we find a very different situation. A recent study provided a survey on religious and spiritual communities in Berlin and its commuter belt, and counted more than 360 of them (Grübel and Rademacher 2003: vi). International migration undoubtedly plays a decisive role in the comparatively high number and plurality of religious orientations found, but so do the so-called new religious movements, or what the authors named "religious trends since the enlightenment" and "postmodern religiosity." The latter comprise at least a third of the counted groups, among them are Bahá'í, Western Buddhists, adherents of the "I AM" and Hare-Krishna movement,[13] Rosicrucians, Freemasons, followers of the Holy Grail movement and of Scientology, Neopagan witches and many more (Grübel and Rademacher 2003: 499–602[14]). Meanwhile, Berlin's rural-urban areas also serve as a ground of religious/spiritual experimentation. In 1991, a small group of men and women set up the "Center for experimental societal organization" close to the German capital. They formed a community that lives together and offers seminars each year in shamanic and tantric practices, earth-based religions and meditation.[15] The center "radiates" back to the city and works as a kind of hub and popular retreat for Berlin's new religious "scene."

Overall, Berlin and its surrounding provide such a myriad of new religious/spiritual groups that, in total, they are "even for insiders [of the "scene"] hardly assessable in detail" (Grübel and Rademacher 2003: 600). They remain rather fluid, barely institutionalized and perhaps "underground" in nature, and thus have developed wide-ranging ways of spiritualizing Berlin and its urbanesque habitats. Take, for example, the group of Neopagan witches, which is growing quickly in the city and comprises approximately 500–600 followers.[16] Their site of worship is "nature," which they perceive as sacred and in opposition to the city, which – according to the witches – has aliened humans from "natural" forces and rhythms and thus from the divine over all (Magliocco 2004; Salomonsen 2002; Hegner 2013). For their rituals, the witches retreat to the lakes and recreational areas at the urban periphery and further to the forests and fields in the commuter zone. At their chosen ritual sites, they draw huge spirals on the ground or erect stone circles, which for them symbolize the sacred circle of life: birth, maturation, death, and rebirth. Through those demarcations of space and through their performances, witches create a *spiritual topography* that traverses Berlin's urbanesque and the urban periphery. Witches spiritualize those areas as well as the city and its inhabitants. For, they believe, the ritual remnants—the spirals, the stone circles—unfold a specific "energy." As urban dwellers, commuters included, pass by those ritual places, they might discover the symbols and thus—the witches are certain—will experience the ritual sites' "sacred power."

As the example of the witches demonstrates, new religious groups shape the cityscape as much as ideas of what the city represents models their religious practice. In this context, Berlin, with its specific religious plurality juxtaposed with its tendency towards secularization, represents a paradigmatic locale of the simultaneity of spiritualizing and de-spiritualizing processes in and around the city.

Spiritualizing the City: its contents and findings

To discuss the various ways cities are nowadays being spiritualized and spatially adapted and designed, we categorized the research results into three distinct themes, which are discussed here in more detail.

Part I of the book explores the "Politics of Religious Plurality and Identity." The two contributions address the political-cultural instrumentalization of religion and the question of how and to what extent governmental institutions and religious organizations have to manage religious expressions within their mandate and to realize practical adaptations in the urban and urbanesque environment. Changes in urban culture, including religious pluralities as well as caesuras in demography, demand that religions creatively rethink established organizational structures and try out new forms of institutionalization and of public visibility, as well as of communication and religious claims.

To that end, Eva Dick and Alexander-Kenneth Nagel compare how during the last decade two German cities—Hamm and Duisburg—created different inter-religious dialogues and "faith governance" to make changes and additions in the cityscape possible and still uphold urban conviviality. As the concept of integration has become a leading policy objective on municipal agendas in Germany and within the recent intercultural paradigm, the governance of religious diversity has been included in many urban integration plans. Dick and Nagel examine how the varied initiatives and networks that developed, always involve the exploration of the margins of religious traditions and trigger new and hybrid religious worldviews (such as, dialogue meetings). Simultaneously, as the authors elaborate, these initiatives reinforce existing mechanisms and tendencies of exclusion. In this context, the study's comparative lens illuminates the extent to which inter-religious initiatives remain context-bound, shaped by the "kind of town" they take place in, that is, its specific history and political culture.

Tricia C. Bruce leads us into a complete different sphere of governance, which is a more one-sided one, feasibly carried out by the Roman Catholic Church through its hierarchical structure. Her research concerns the extensive redistribution of church buildings connected to full new community building in the American city of St. Louis and its agglomeration. Bruce explains that the Catholic authorities had to exchange the logic of propinquity of previous church communities for a diasporic community, with members spread all over the outskirts of the agglomeration. Hence, parishes are no longer defined according to territory, but according to special purposes and different needs of Catholic subcultures varying by ethnicity, by worship style, and by primary mission. Only through this fundamental innovation of organizational structures could existing church buildings—magnificently built and central symbols of St. Louis's cultural image of being the "The Rome of the West"—be preserved and Catholic presence in the city and its greater area maintained. For the new communities, policies on architectural cultural heritage and the collective memory of the religious urban past became important, bringing in new forms of spiritual tourism as incentive for the urbanesque.

Part II of the book is entitled "Producing and Negotiating Religious Space" and discusses the various ways in which new religious movements generate presence in the city realm and beyond, by creating their own space and appropriating particular places and spaces, hardly without any governmental interference. The question of (in)visibility also relates to the issue of self-representation of religious movements, its aesthetics and urban spreading of their following. In this context, the section provides several perspectives. It shows how neglected, abandoned parts of the urban and the urbanesque are appropriated by religious groups and bring again urban deserts back to life. It further illuminates how religious practices can generate a specific perception or (religious) image of a city as a whole. Part II also draws attention to new sacral architecture that is a bricolage of different religious styles, and helps to shape the cityscape. By concentrating on innovative trends within Islam, the final chapter in this section looks at the creation of religious spaces and topographies within the urban and urbanesque habitat that remains discursively highly contested.

We introduce this section with Gertrud Hüwelmeier's contribution, in which she explores how religious place-making emerges from the ways in which migrants transport and introduce religious ideas, practices and sacred objects from one place to another, while simultaneously changing or redefining ideas about belief, ritual, locality and sacred space in the urban context. She demonstrates in "Praying in Berlin's 'Asiatown'" how Asian religions have created through their migrants a kind of independent town as an expression of "recycled" place-making in former industrial and business zones. The established product markets—multi-ethnic bazaars—are fused with religious practices. Prayers and attendance at set-up altars are easily integrated into the daily work routine. Lately and additionally, sacred spaces such as a pagoda and a mosque have been built inside and outside of the buildings. Generally, Hüwelmeier draws attention to the intense interplay of local ties and global networks at these single locales and thus shows that the markets in their suburban and outdated industrial zones have become keys to religious transnationalism, whether or not in combination with market and trading activities.

Anna Niedźwiedź combines long-term fieldwork with personal observations and notes, and analyzes how the inhabitants of Kraków, in the middle of a process of general religious decline in Poland were able to re-spiritualize Kraków's cityscape in a mixed way. The global admiration and veneration of Pope John-Paul II became "nationalized" in Poland, and due to his Polish and Kraków upbringing, his cultus was specifically inscribed into its cityscape. At first instance, this came about when people began to make grass-roots street shrines in downtown streets, which led to a new, institutionalized popular devotional practice: the cultus of the "Papal Window." This public urban spiritualization facilitated again the creation and the success of new shrines of pilgrimage created in one of the redeveloped industrial areas outside of town, which were related to the Pope's life and to his favored devotion to the Divine Mercy. Those forms of religious practice and place-making, again, reproduce and generate Kraków's cultural and mental image of being the "Pope's city": an image that is an important resource of identification for Poles and Catholics all over the world.

Mariske Westendorp researches one of the densely populated cityscapes in the world. The density of Hong Kong's built environment is reflected in the meanings that individual religious practitioners ascribe to religious buildings as material mediators between themselves and the divine. She explores, based on their narratives, how Catholic and Buddhist practitioners in Hong Kong relate to specific religious buildings—churches and temples—and reveals that the location and the aesthetics of church buildings are not the most critical media for religious practice. They are taken as "peaceful and quiet shelters" for mere personal devotional practices within the turbulent environment of Hong Kong. As Westendorp moves on in her analysis, she points to Hong Kong's rather unique situation, as it is bounded by the ocean and the border with mainland China that cannot be geographically extended. Since land is thus not only scarce but expensive, particularly in the (ever-decreasing) urbanesque areas that allow a retreat from Hong Kong's hectic (that is, the New Territories), new religious centers do not spread in Hong Kong and beyond. Instead, they are concentrated in the inner city and are increasingly established in already existing residential, industrial, or commercial buildings. In a way, one can conclude, Hong Kong's situation as a geographical and cultural-political island, intensifies the spiritualization of central districts of this Asian megacity.

Claire Dwyer explains in her contribution on the spiritualizing of the suburbs how new religious architecture in suburban London and Vancouver brings in the sensation of the visual aesthetics. She explores the spiritualizing of the city by considering the transnational geographies of new forms of religious architecture built in the suburbs by migrant faith communities. The examples explored are a Tibetan monastery, in Richmond, Vancouver, a Jain temple in Potters Bar, London, and a Muslim center under construction in Harrow, near London. Dwyer argues that those buildings are examples of new forms of hybrid religious architecture creating new suburban sacred spatiality, subsequently also transforming the existing meanings of suburban space.

Synnøve Bendixsen's contribution brings us back to the multi-religious cityscape of Berlin. She addresses the spiritualizing of Berlin through a discussion of how the image of a multicultural Berlin becomes part of how Muslim youth create a "fun Islam." Fun Islam, an emic term, actually refers to a conservative form of Islam. It puts forward the idea that pious strictness can be "fun," without segregating oneself from society; it thus has to be read as an (urban) response to the highly contested discursive field of the role of Islam in the "Western" world. By focusing on one of the most active Muslim youth organizations in the city, Bendixsen shows the complex ways young Muslims create spaces for their particular concept of Islam and negotiate religious presence in Berlin. Venturing into the city, as Bendixsen reveals, they aim to be perceived publicly simultaneously as Muslims and as "Germans" and "Berliners," who synthesize their piety with an openness towards a "multicultural" way of life in the city and its urbanesque.

Part III is entitled "The Agency of the Body and the Senses in Spiritualized Practices." This final part concentrates on the question of how cities and the urbanesque facilitate spiritual and religious movements in relation to corporality, emotions and the senses. The three contributions in this section focus particularly on the role of music and how listening to music, and dancing, singing and performing, through bodily and emotional engagement, are involved in establishing new religious movements with a need for dedicated religious spaces in the urban and urbanesque.

To that end Sarah Pike points out how trance or ecstatic dance is to be seen as a prayer to the self. Focusing on "Five Rhythms dance churches" in the San Francisco Bay Area as well as in other urban and urbanesque regions in the US, she depicts how the participants are "sweating their prayers," in reference to one's "Rhythms of the Soul." Although they claim their dance worship to be an "antidote of city life," dancers nevertheless bring their urban experiences in with them and take experiences forged in dance sanctuaries, again, back on to city streets and into other urban spaces. Pike examines this specific interplay of escape from and immersion in urban life that characterizes the spread of this New Age form of ecstatic dance. In this context, she draws particular attention to the places where those dances take place—old warehouses, school gyms and community centers—and shows how dancers transform these spaces into temporary temples of movement. This form of spiritualizing space, as Pike examines, is a complex attempt to separate the sacred from the city and its urbanesque. Since those buildings carry traces of their other lives, it requires selective remembering of particular urban pasts and forgetting of others.

Daniel Wojcik and Peter Jan Margry present another sensorial expression of spiritualization of the urban. Their research describes how the performance and especially the perception of John Coltrane's jazz music, has changed due to gentrification processes in the Western Addition and Fillmore districts of San Francisco. After a governmental order to incubate the free-floating movement around a Godly Coltrane into a formal Orthodox Church organization and an ejection of the Coltrane devotees of their initial "Harlem-of-the-West" locality due to rising property prices, the group moved to a different, completely renovated part of town. This resulted in a shift in its audience, which changed from a group of predominantly local African American jazz enthusiasts to a more "whitened" audience arriving from all over the country and the world—an audience apparently more interested in the stand-alone sacred and transformative power of the sound of the album *A Love Supreme* than in the offered services of the African Orthodox Church and its members.

Raphaela von Weichs also addresses dance music and prayer: "Singing is prayer two times" is the saying in Cameroon. She brings up a transnational perspective on "religious music," musical performance and urban religiosity in Cameroon and Switzerland, and examines the performance and politics of "religious (Christian) music" in urban Cameroon and the incorporation of Cameroonian choral music into "African pilgrimages" in Switzerland. Music, especially urban "religious music" in Cameroon, von Weichs reveals, has brought new dynamics into modern city life and its urban religious practices. Its incorporative and transformative power, however, is particularly tangible in situations of translocal and transnational migration. This is especially the case when place-making is difficult to achieve and migrants live dispersed in cities and urbanesque habitats. In such processes of localization, the semi-rural sites (ancient Catholic abbeys) of newly invented Swiss African pilgrimages turn into a stage for (in the Swiss cultural context) innovative musical and religious practices, such as the performance of Cameroonian choir music. It becomes obvious that religious transformation is not confined to the city but results from a constant interplay between urban and rural places.

To conclude this introduction to these intriguing case studies, the editors would like to affirm that modern metropolises are spiritualized—they always were, in historically very different ways. Presently, we are witnessing a "spiritual boom"-time in cities and their urbanesque habitats: religions, religiosities and spiritualities pluralize and forcefully create urban public spaces or generate socio-cultural niches for their worldviews and practices. It is a development which is highly interrelated with the success of cities themselves as desirable life-worlds, identity-laboratories and (imagined) sustainable places of prosperity. An intense mutual agency and resilience makes religious practice thrive. These dynamics not only count for the "classic" urban space but need to be taken into account beyond the traditional borders of academic understanding. The migration of people all over the world brings new focal points for religious experiments, encounters and mixtures, and as people react strongly to the extended cityscape we have taken for this book: not a static physically built one, but the one related to people's practices, agency and behavior. Such a perspective clearly shows that religious practice is no longer related to formal sacral buildings or exclusive sacral grounds but finds its extended habitat where the agency of older buildings and former industrial and business zones create new chances, and where new environments and emergent spaces nurture the spiritual impulse, and can adapt to the possibilities there.

In summary, this book exemplifies how urban space, newly defined in the broader culturally lived way, is the focal point of present-day religious practice and shows how this basic human expression is transformed by the urban domain, and by its power of belief, transforms urban culture.

Notes

1 See https://sundayassembly.com; in 2014, the article "the" was deleted from the official naming (originally "The Sunday Assembly"); also the wording of their mission and vision have slightly been changed since.
2 Website statements (https://sundayassembly.com/about/), accessed on respectively April 3, 2014 and August 18, 2015. As the movement is quite recent and is quickly developing websites, texts are due to change; see, for Amsterdam, http://amsterdam.sundayassembly.com.
3 Fieldnotes, October 25, 2015.
4 https://www.sundayassembly.com/story; accessed March 5, 2016.
5 https://www.cnu.org/who-we-are/charter-new-urbanism; accessed March 5, 2016.
6 http://globalprayer.info; accessed March 5, 2016.
7 At a symposium called "City as Context" in 1973, held by the American Anthropological Association in New Orleans, the cultural anthropologist Jack Rollwagen already suggested the method of comparative studies in two or more cities as a useful means to understand the cultural distinctiveness of urban locales: Rollwagen 1975: 54–6.
8 Also in implicit ways in "religion"-void suburban(esque) areas, cf. Schippers 2015.
9 Cf. Patrick Meershoek, "Bidden in de parkeergarage," in *Het Parool*, May 18, 2004.
10 Also in other cities, plans exist for combined buildings, like Berlin's "House of One," which houses a church, a mosque and a synagogue under one roof: *The Guardian*, June 25, 2014.
11 Corrie Verkerk, "De heilige geest in alle stadsdelen," in *Het Parool*, May 24, 2002; Remco Tomesen, "De nieuwe buurtkerk doet het goed," in: *Het Parool*, July 31, 2012; Sjaak van de Groep, "Het is weer volle bak in de vergrijsde stadskerk," in *NRC-Handelsblad*, March 16, 2012.
12 Amsterdam's statistics are even lower: only 10 percent of its inhabitants belong to the Catholic or Protestant Church: http://www.ois.amsterdam.nl/assets/pdfs/2014_religie_in_amsterdam.pdf; accessed March 24, 2016. To the extent to which Berlin could be called an "exceptional case" for Germany, see Thomas Großbölting's study on Germany's "religious situation": Großbölting 2013.
13 Grübel and Rademacher categorize the "Hare-Krishna Movement," also known as the "International Society for Krishna Consciousness" (ISKCON), in a slightly different way as a "new community in the tradition of Indian religions": Grübel and Rademacher 2003: 394–5.
14 Interview with representatives of the "Berlin office for sect-related issues," January 14, 2009. The office monitors religious/spiritual groups outside the dominant monotheistic religions. The two main churches in Germany established similar offices; however, their "labels" sound less ideological, for example, the Protestant Central Office for World-View-Questions (*Evangelische Zentralstelle für Weltanschauungsfragen*).
15 http://www.zegg.de/ (main homepage); accessed March 5, 2016. Earth-based religions are religions that worship the earth and nature as immanently sacred and advocate ecological activism.
16 It is difficult to provide rather reliable statistics. According to the religious studies scholar, Reena Perschke (2003), approximately 350–400 witches live in the city. However, the witches themselves say those numbers are far too low and don't take recent developments into account. Estimates now range from 500–700.

References

Al Sayyad, Nezar, and Mejgan Massoumi (eds) (2011), *The Fundamentalist City? Religiosity and the Remaking of Urban Space*. London and New York: Routledge.
Bailey, Edward (1999), *Implicit Religion: An Introduction*. London: Middlesex University Press.

Beaumont, Justin, and Christopher R. Baker (2011) (eds), *Postsecular Cities: Space, Theory and Practice*. London and New York: Continuum International Publishing Group.

Becci, Irene, Marian Burchardt, and José Casanova (eds) 2013, *Topographies of Faith. Religion in Urban Spaces*. Leiden: Brill.

Becker, Jochen, Katrin Klingan, Stephan Lanz, and Kathrin Wildner (eds) (2014), *Global Prayers: Contemporary Manifestations of the Religious in the City*. Berlin: Lars Müller.

Berger, Peter L., (1967), *The Sacred Canopy: Elements of a Sociological Theory of Religion*. Garden City, NY: Doubleday.

Berger, Peter L. (ed.) (1999), *The Desecularization of the World: Resurgent Religion and World Politics*. Grand Rapids, MI: Eerdmans Publishing.

Berger, Peter L. (2001), Postscript, in Linda Woodhead, Paul Heelas and David Martin (eds), *Peter Berger and the Study of Religion*. London: Routledge, 189–98.

Burchardt, Marian, and Irene Becci (2013), Introduction: Religion Takes Place: Producing Urban Locality, in Irene Becci, Marian Burchardt and José Casanova (eds), *Topographies of Faith. Religion in Urban Spaces*. Leiden: Brill.

Burgess, Ernest (1925), The Growth of the City: An Introduction to a Research Project, in Robert E. Park and Ernest Burgess (eds), *The City*. Chicago, IL: Chicago University Press, 47–62.

Carrette, Jeremy, and Richard King (2005), *$elling Spirituality. The Silent Takeover of Religion*. London: Routledge.

Cimino, Richard et al. (eds) (2013), *Ecologies of Faith in New York City. The Evolution of Religious Institutions*. Bloomington: Indiana University Press.

Clapp, James A. (2012), *The City: A Dictionary of Quotable Thoughts on Cities and Urban Life*. New Brunswick, NJ: Transaction Publishers.

Coleman, Simon (2013), Scholarly Languages II: Anthropology, Religion and Migration, in Jane Garnett and Alana Harris (eds), *Rescripting Religion in the City. Migration and Religious Identity in the Modern Metropolis*. Burlington, VT: Ashgate, 45–50.

Cox, Harvey (1984), *Religion in the Secular City: Toward a Postmodern Theology*. New York: Simon and Schuster.

Cox, Harvey (2013; 1st edn. 1965), *The Secular City: Secularization and Urbanization in Theological Perspective*. With a new introduction by the author. Princeton, NJ: Princeton University Press.

Day, Katie (2014), *Faith on the Avenue. Religion on a City Street*. Oxford and New York: Oxford University Press.

Fuller, Robert C. (2001), *Spiritual, But Not Religious. Understanding Unchurched America*. Oxford: Oxford University Press.

Gómez, Liliana, and Walther van Herck (2013), *The Sacred in the City*. London: Continuum.

Großbölting, Thomas (2013), *Der verlorene Himmel. Glaube in Deutschland seit 1945*. Göttingen: Vandenhoeck & Ruprecht.

Grübel, Nils, and Stefan Rademacher (2003), *Religion in Berlin. Ein Handbuch*. Berlin: Weissenseeverlag.

Hanegraaff, Wouter J. (1996), *New Age Religion and Western Culture. Esotericism in the Mirror of Secular Thought*. Leiden: Brill.

Hannerz, Ulf (1980), *Exploring the City: Inquiries Toward an Urban Anthropology*. New York: Columbia University Press.

Heelas, Paul, and Linda Woodhead (2005), *The Spiritual Revolution. Why Religion is Giving Way to Spirituality*. Oxford: Blackwell.

Hegner, Victoria (2008). *Gelebte Selbstbilder. Gemeinden russisch-jüdischer Migranten in Chicago und Berlin.* Frankfurt am Main: Campus Verlag.

Hegner, Victoria (2013). Hex and the City. Neopagan Witchcraft and the Urban Imaginary in Berlin, *Ethnologia Europaea*, 43(1): 88–97.

Jacobs, Jane (1992), *The Death and Life of Great American Cities.* New York: Vintage Books.

Kaschuba, Wolfgang (2015), Vom Wissen der Städte. Urbane Räume als Labore der Zivilgesellschaft, in Wolfgang Kaschuba, Dominik Kleinen and Cornelia Kühn (eds), *Urbane Aushandlungen. Die Stadt als Aktionsraum.* Berlin: Panama Verlag, 13–29.

Katz, Peter (1994), *The New Urbanism: Toward an Architecture of Community.* New York: McGraw Hill.

Kersting, Franz-Werner, and Clemens Zimmermann (2015), Stadt-Land-Beziehungen. Geschichts- und kulturwissenschaftliche Perspektiven, in Franz-Werner Kersting and Clemens Zimmermann (eds), *Stadt-Land-Beziehungen. Geschichts- und kulturwissenschaftliche Perspektiven.* Paderborn: Schöningh, 9–34.

Knoblauch, Hubert (2009), *Populäre Religion. Auf dem Weg in eine spirituelle Gesellschaft.* Frankfurt am Main: Campus.

Knott, Kim (2005), *The Location of Religion: A Spatial Analysis.* London: Equinox Publishers.

Korff, Gottfried (1985), Mentalität und Kommunikation in der Großstadt. Berliner Notizen zur 'inneren, Urbanisierung, in Theodor Kohlmann, and Hermann Bausinger (eds), *Großstadt. Aspekte empirischer Kulturforschung. 24. Deutscher Volkskunde-Kongreß in Berlin.* Berlin: Staatliche Museen Preussischer Kulturbesitz, 343–61.

Lanz, Stephan (2014), Assembling Global Prayers in the City: An Attempt to Repopulate Urban Theory with Religion, in Jochen Becker, Katrin Klingan, Stephan Lanz and Kathrin Wildner (eds), *Global Prayers: Contemporary Manifestations of the Religious in the City.* Berlin: Lars Müller, 17–45.

Lee, Martyn (1997), Relocating Location: Cultural Geography. The Specificity of Place and the City Habitus, in Jim McGuigan (ed.), *Cultural Methodologies.* London: Sage, 126–41.

Leezenberg, Michiel (2010), How Ethnocentric is the Concept of the Postsecular, in Arie Molendijk, Justin Beaumont and Christoph Jedan (eds), *The Religious, the Political and the Urban.* Leiden and Boston, MA: Brill, 91–112.

Lindner, Rolf (1993), Berlin – Zone in Transition, *Anthropological Journal on European Cultures*, 2(2): 99–111.

Lindner, Rolf (1996), Perspektiven der Stadtethnologie, *Historische Anthropologie*, 5(2), 319–28.

Lindner, Rolf (2003), Der Habitus der Stadt – ein kulturgeographischer Versuch, *Petermanns Geographische Mitteilungen*, 147(2): 46–53.

Lindner, Rolf (2006), The Imaginary of the City, in Günter H. Lenz, Friedrich Ulfers and Antje Dallmann (eds), *Toward a New Metropolitanism. Reconstituting Public Culture, Urban Citizenship, and the Multicultural Imaginary in New York and Berlin.* Heidelberg: Winter, 209–16.

Livezey, Lowell W. (ed.) (2000), *Public Religion and Urban Transformation: Faith in the City.* New York: New York University Press.

Luckmann, Thomas (1967), *The Invisible Religion. The Problem of Religion in Modern Society.* New York: Macmillan.

MacKian, Sara (2012), *Everyday Spirituality. Social and Spatial Worlds of Enchantment.* Basingstoke: Palgrave Macmillan.

Magliocco, Sabina (2004), *Witching Culture. Folkore and Neo-Paganism in America.* Philadelphia: University of Pennsylvania Press.

Margry, Peter Jan (2012), European Religious Fragmentation and the Rise of Civil Religion, in Ullrich Kockel, Jonas Frykman and Máiréad Nic Craith (eds), *A Companion to the Anthropology of Europe*. Malden: Wiley-Blackwell, 275–94.

McLeod, Hugh (1996), *Piety and Poverty: Working-Class Religion in Berlin, London, and New York, 1870–1914*. New York: Holmes & Meier.

metroZones e.V. (eds) (2011), *Urban Prayers. Neue religiöse Bewegungen in der globalen Stadt*. Berlin and Hamburg: Assoziation A.

Mohrmann, Ruth-E. (ed.) (2010), *Alternative Spiritualität heute*. Münster: Waxmann.

Oomen, Mar, and Jos Palm (1994), *Geloven in de Bijlmer. Over de rol van religieuze groeperingen*. Amsterdam: Het Spinhuis.

Orsi, Robert A. (ed.) (1999), *Gods of the City. Religion and the American Urban Landscape*. Bloomington: Indiana University Press.

Partridge, Christopher (2004), *The Re-enchantment of the West. Alternative Spiritualities, Sacralization, Popular Culture and Occulture*, 2 vols. London: T. & T. Clark International.

Perschke, Rena (2003), Neuheidnisches Hexentum. Wicca. Pagan. Freifliegende, in Nils Grübel and Stefan Rademacher (eds), *Religion in Berlin. Ein Handbuch*. Berlin: Weissenseeverlag, 525–9.

Pinxten, Rik and Lisa Dikomitis (eds) (2012), *When God Comes to Town. Religious Traditions in Urban Contexts*. New York: Berghahn Books.

Porter, Libby, and Kate Shaw (eds) (2009), *Whose Urban Renaissance? An International Comparison of Urban Regeneration Strategies*. London and New York: Routledge.

Punter, John (2010) (ed.), *Urban Design and the British Urban Renaissance*. London and New York: Routledge.

Rogier, Louis J. (1945), *Geschiedenis van het katholicisme in Noord-Nederland in de 16e en de 17e eeuw*. Amsterdam: Urbi et Orbi.

Rollwagen, Jack R. (1975), The City as Context: The Puerto Ricans of Rochester, *Urban Anthropology*, 4(1): 53–9.

Salomonsen, Jone (2002), *Enchanted Feminism. Ritual, Gender and Divinity among the Reclaiming Witches of San Francisco.* London and New York: Routledge.

Saraiva, Clara, Peter Jan Margry, Lionel Obadia, Kinga Povedák and José Mapril (eds) (2016), *Experiencing Religion. New Approaches to Personal Religiosity*. Berlin: Lit-Verlag.

Schippers, Inez (2015), *Sacred Places in the Suburbs. Casual sacrality in the Dutch VINEX-district Leidsche Rijn*. Amsterdam: Institute for Ritual and Liturgical Studies.

Schmidt-Lauber, Brigitta (2010), Urbanes Leben in der Mittelstadt. Kulturwissenschaftliche Annäherungen an ein interdisziplinäres Forschungsfeld, in Brigitta Schmidt-Lauber (ed.) 2010. *Mittelstadt: Urbanes Leben jenseits der Metropole*. Frankfurt: Campus, 11–36.

Smith, Michael P. (2002), Power in Place: Re-theorizing the Local and the Global, in John Eade and Christopher Mele (eds), *Understanding the City: Contemporary and Future Perspectives*. Oxford: Blackwell Publishing, 109–30.

Statistisches Jahrbuch Berlin 2005. Berlin: Kulturbuch-Verlag.

Statistisches Jahrbuch Berlin-Brandenburg 2015. Berlin: Kulturbuch-Verlag.

Stoller, Paul (1997), *Sensuous Scholarship*. Philadelphia: University of Pennsylvania Press.

Strauss, Anselm and Richard R. Wohl (1958), Symbolic Representation and the Urban Milieu, *American Journal of Sociology*, 63(5): 523–32.

Strong, Josiah (1885), *Our Country. Its Possible Future and Present Crisis*. New York: The American Home Missionary Society.

Suttles, Gerald (1984), The Cumulative Texture of Local Urban Culture, *American Journal of Sociology*, 90(2): 283–304.

Urban Task Force (1999), *Towards an Urban Renaissance.* London: Spon.

Van der Valk, Leendert (2014), Afrikaanse Spirits, spirituele Vieringen [African Spirits; Spiritual Celebrations], *NRC-Handelsblad*, January 17.

Part I

The politics of religious plurality and identity

2 Local interfaith networks in urban integration politics

Religious communities between innovation and cooptation

Eva Dick and Alexander-Kenneth Nagel

Introduction

Over the last decade, integration has become a leading policy objective in Germany (Bommes 2007: 3). This is manifested in a variety of integration measures taken both on the national as well as on the local level. Regarding the national level, in the year 2005, after decades of neglecting the de facto situation of being an immigration country, Germany has introduced an immigration law that also encompasses structures and strategies for integration. High-level policy events such as the National Integration Summit conducted annually since 2006, the German Islam Conference (also conducted since 2006) and the formulation of a National Integration Plan in 2007[1] indicate that the issue of integration is high on the agenda of national German politics and that faith and religion have been strong incentives to that end, especially due to the increase of Islamic extremism. On the municipal level, where integration challenges have always manifested themselves more directly, most cities have now formulated local integration or intercultural plans.

Over the last few years, German normative ideas on and concepts of integration have changed considerably. In the context of the growing role of multiculturalism in the integration discourse (Leggewie 2009: 593), emphasis is placed on strategies for improving mutual understanding and conviviality, implying an increased attention towards different people's cultures and religions. Particularly in the 2000s, following 9/11 and subsequent violent attacks, interreligious dialogue and understanding were envisioned by politicians as key instruments to manage intercultural conflicts and Islamist radicalism from the local to global levels.

While changing integration paradigms play an important role, this chapter argues that in Germany, as in other Western societies, the governance of religious diversity within urban integration policies is significantly shaped by two further factors. First, religious pluralization through migration and related claims on urban conviviality and space particularly manifests itself in specific urban "*geographies* of post-secular activity," for example, in ethnically diverse neighborhoods (Cloke and Beaumont 2013: 32, emphasis in original; Gale 2008). And second, new forms of urban governance in which religious representatives have become relevant partners in the negotiation and development of urban policies have emerged, notably with regards to integration and diversity management. We may thus understand interfaith activities and networks as new forms of "embodiment" of the religious/spiritual in today's postmodern cityscapes shaped by migration and multicultural encounter (Sandercock 1998, 2003; Sandercock and Senbel 2011). In this chapter, we refer to the term "religion" to denote agency in the name of a distinct religious tradition (such as Islam) or denomination (such as Alevites). In addition, our understanding of spirituality involves a broader notion of transcendent experiences with a tendency to blur or to transgress traditional religious boundaries.

This chapter is organized as follows. The subsequent, second, section discusses concepts of integration and governance basic for framing the debate about religious diversity in the urban context. In the third section, the data and methods underlying our analysis are described and rationales for choosing our city cases are explained. In the empirical part of this paper (the fourth and fifth sections), we explore how the governance of religious diversity is put into practice on the municipal level against the backdrop of changing integration paradigms and participatory forms of urban governance. This will be done by analyzing the significance of religious diversity in municipal integration programs and assessing the involvement of public stakeholders in the broad field of local interreligious activities, in the form of a preliminary typology. We conclude with assessing the patterns of official interreligious cooperation in the two cities in a comparative way and spelling out new perspectives for urban integration by means of the governance of religious diversity.

The enhanced role of religion in urban space

Over the last few years, a number of studies have pointed to the increasing relevance of religion in the public sphere,[2] notably in urban areas. These studies reflect the enhanced involvement of faith-based communities in urban integration and cohesion efforts. As these efforts constitute joint projects of regulation between state and societal actors, we conceive of them as governance arrangements (Mayntz and Scharpf 1995).

Since the 1970s and 1980s, when municipalities in Germany began to formulate strategies for better integration of the immigrant population, normative concepts underlying them have evolved considerably. At first, strategies were aimed at countering integration deficits of labor migrants and their children. These so-called "compensating integration measures" (Baraulina 2007) focused on enhancing immigrants" employment, education and housing. In the course of the last few decades, they were complemented by other approaches placing higher emphasis on migrants" assumed specific competences. These so-called "resource-oriented" integration measures emphasize the two-sidedness of integration processes as well as the (inter) cultural or value-related dimension of integration (Baraulina 2007: 30).

Against this background, from the national down to the local level, the intercultural discourse has become dominant in describing migration-related policies.

It facilitates the entry of religion in integration policies insofar as religion is conceived of as a basic element of people's "culture" (Tezcan 2006; Nagel 2012). Until the beginning of the 2000s, interreligious dialogue was particularly led by civil society and centered on Christian-Muslim relations. Today, in contrast, organized interreligious encounters involve a larger spectrum of religious organizations and include public sector representatives. They consider it as a key channel and platform for the promotion of intercultural understanding and community cohesion (Klinkhammer et al. 2011; Tezcan 2006; Kufen 2008). This broader participation is all the more the case due to a perceived religious dimension of social relations and of many global and local conflicts (Leggewie 2009: 593).

Cities as main destinations of international migration and clusters of religious pluralization are privileged sites in which religious or spiritual claims on participation and space are put forward (Griera and Forteza 2011: 113, 117; Leggewie 2009: 600–602, Gale 2008). More concretely, these claims are typically negotiated in or with respect to certain urban localities, such as dilapidated and/or ethnically mixed neighborhoods or undetermined areas in terms of land usage (Dinham et al. 2009: 6; Weller 2009: 70.).[3] It is also often in or with reference to these spaces and their inhabitants that "post-secular" alliances between the state and religious groups are forged (Cloke and Beaumont 2013: 28; metroZones 2011; Beaumont and Baker 2011). Such alliances may be considered an "embodiment" of the postmodern urban condition heavily shaped by international migration and intercultural encounter (Sandercock 1998, 2003; Sandercock and Senbel 2011: 88).

Religious actors and/or interfaith bodies are increasingly regarded as important civil society participants in urban governance arrangements (Dinham and Lowndes 2009: 5). In general terms, such cooperative forms of governance accommodate claims for enhanced civic inclusion and "empowerment" in public policies, yet they also need to be (critically) viewed against the background of neoliberal agendas to reduce governmental responsibility in society (Healey 2007: 18; metroZones 2011: 19; Eade 2011). With respect to religious or interreligious organizations, the literature posits at least two further factors responsible for their growing involvement in urban governance arrangements: first, the enhanced capability of religious players to advance their stake, due to an increased institutional formalization and sociopolitical positioning (Penta and Schiffauer 2011: 252; Cloke and Beaumont 2013: 31), and second, the "religious illiteracy" of policy makers and planners in Western secularist[4] metropolises (Griera and Forteza, 2011: 124–5; Leggewie 2009: 597).

Both factors lead to (inter)religious groups fulfilling an important broker and communicator function between municipal stakeholders and urban immigrant communities (many of them identifying along religious lines), and between potentially conflicting ethnocultural or religious groups (Baker 2008: 7–8; Dinham and Lowndes 2009: 6).

We are particularly interested in analyzing the outcomes of the involvement of (inter)faith networks in what may be considered a continuum from religious innovation to instrumentalization. On the one end of this spectrum, the involvement of (inter)religious organizations in public-private governance arrangements may develop at the cost of their religious agendas (Dinham et al. 2009: 5). On the other end, public-private partnerships between state representatives and religious actors may result in innovative and more inclusive state-religion relationships, for example, in the sense of effectively accommodating new and more diverse spiritual ways of being and acting (Weller 2009: 78), which Cloke and Beaumont consider as positive co-governance in which interests of both parties end up being met.

Research context and methodology

This chapter grows out of an interdisciplinary research project conducted in Germany between January 2012 and August 2013.[5] Against the notion of a reappearance of religion in today's public and urban space, the project's objective was to analyze the transforming relationship between state and religious actors by focusing on the governance of interreligious contacts and activities in selected urban areas.

Methods of data generation comprise semi-structured individual interviews with representatives of the city and religious initiatives, focus group discussions, document analysis and participant observation. The term "city representatives" includes members of the local government and administration of two selected cities, whereas "religious representatives" refers to persons involved with religious communities or interreligious activities at the local level. The analysis of the role of interreligious activities in urban integration plans in the fourth and fifth sections of this chapter is primarily based on the present urban integration concepts of the two cities. Further sources were semi-structured interviews with city stakeholders and observation notes. The interviews were analyzed by means of a structured content analysis, in which key concepts extracted out of the literature (modes of state-religious interaction, inclusion/exclusion of religious representatives, representation) were used as starting points of the analysis.

Our analysis is based on two city case studies, namely, Hamm and Duisburg. Both cities are located in the Ruhr Area, the largest urban agglomeration area within Germany and among the regions most strongly influenced by the labor immigration during the second half of the twentieth century (the "guest workers"). Both religious diversity and the density of religious organizations are particularly elevated in the Ruhr Area, in comparison to other regions of the State of North Rhine-Westphalia, due to its highly urbanized character (Krech 2008: 34–6). In contrast, the proportion of people affiliated to religious organizations is comparatively lower in cities than it is in rural areas (Krech 2008: 38). These space-related tendencies are also corroborated by data from other German regions (Leggewie 2009: 595). Socio-economically, the Ruhr Area has undergone significant transformation: While it was the center of the coal and steel industry in Germany for more than 150 years, most of the mines and factories have now been closed down. This has made it rather easy for religious communities to find new habitats in the former industrial zones.

Although both Hamm and Duisburg exhibit a high degree of religious diversity and are confronted with similar socio-economic challenges of post-industrial transformation, they differ significantly regarding their size and political shape: Duisburg consists of approximately 492,000 inhabitants and has been governed by the Social Democratic Party (one of the main left-of-center parties in Germany) ever since the end of World War II, whereas Hamm has some 182,000 inhabitants and has been governed by the Christian Democratic Union (the main right-of-center party in Germany) for the last two decades.

In the following two sections we will discuss preliminary results regarding the role of interreligious activities in municipal integration agendas and the actual participation of urban authorities in interreligious activities on the ground.

Interreligious activities in integration policies in Hamm and Duisburg

Like other urban areas in Germany, in the past ten years the cities of Duisburg and Hamm have placed a much greater policy focus on the integration of immigrants than they had previously. In this section, we assess key features of the most recent integration concepts or plans of the two cities and the importance and role of religion and interreligious activities in these plans (see Table 2.1).

With respect to the goals and definitions of integration, the integration plans in both cities reflect nation-wide trends, in that integration is considered a cross-cutting issue involving a large variety of dimensions and municipal sectors.[6] The intercultural orientation is highlighted in both plans; in Duisburg it is even stated as the main goal of integration policies and in Hamm it is considered a "strategic concept" (Integration concept of Duisburg: 8, Hamm: section 7).

The same can be said regarding the key principles underlying integration strategies: Resource and process orientation as well as intercultural understanding are common denominators in both plans, similar to what has been stated with respect to other German cities (Baraulina 2007). Participation and dialogue figure as key instruments, for example, during the formulation of integration plans.

The focal areas of intervention in both Duisburg and Hamm reflect a blend of measures to ease so-called "structural integration," such as through employment and skills promotion as well as strategies promoting identification and/or social integration through the arts, cultural exchange and intercultural dialogue (Esser 1980).[7] Further, the interviewed stakeholders express awareness of the importance of the "software" dimensions of integration, as pointed out by two administration officials in Hamm: "[We] also need to offer culture-specific services, meaning we also need to sort of account for these kinds of [cultural] needs" (H01S). And with respect to construction in the public park "Gates of the World Religions," ongoing at the time of our interview, an interviewed official from the city's planning office highlights the interaction of religion with urban material space:

> [This is because] religion demands space in the true sense of the word. This place is an actual space of spirituality, also in a metaphorical sense. You also need to provide [physical] space for the exercise of everyday spirituality. And the exercise of religion forms part of that, including a minaret, church bells on Sundays or for the matter, a temple march . . . of the other Hindu community. This belongs to everyday life. (H08S)

Table 2.1 Objectives, fields of action and the role of religion in integration concepts in Duisburg and Hamm

Cities and plan	*Goals of integration policies and definition*	*principles of integration*	*fields of intervention*	*Role of religion/ interreligious activities*
Duisburg Concept of Integration (2010)	Inter-cultural orientation of social structures, administration and urban entities Definition: Integration as cross-cutting issue (including spatial, temporal, relational, self-identifying, cultural, institutional dimensions also related to the "majority population"	– Resource instead of deficit orientation – Integration through dialogue and participation – Individual integration instead of "ethnicization" – Common urban identity: "we are you" – Integration in the neighborhood – Lived inter-culturality Also: Integration as a process	Political participation and civic engagement; Education, intercultural learning, training and employment; Intercultural orientation of the administration; Art, culture and intercultural dialogue; urban development and socio-spatial integration	– No separate field of intervention in integration plan – Mentioned as one among eight focal fields of action of council of integration (local representative body elected by foreign citizens) – Intercultural dialogue to be guided by "*a secular culture providing the framework for all activities, including those of a religious nature*" (20)
Hamm Integration Report (*Beschlussvorlage* for urban council) of the City of Hamm for the Year 2008 (printed version)	Capacitation of migrants to gain security in all aspects of social life whilst being economically and socially successful. Reduction of costs of disintegrative	– Holistic approach, but language and education are considered "keys" to integration – Non-ethnicization and culturalization of existing integration deficits	Support of language and general orientation; Support of early-childhood, pre-school and school education and training; Support of participation and reduction of barriers for youth in professional	– No separate field of intervention in integration plans – In intervention field "neighborhood-based work and civic engagement": Religious organizations mentioned as important actors on the neighborhood level

processes in a city highly shaped by migration. Definition: Integration as a cross-cutting and community issue	– Recognition of potentials and contributions of migrants for urban society (mobility, language, social bonding and intercultural literacy), in view of increasing internationalization and globalization	formation and employment; Intercultural opening of administration as cross-cutting task; Offers for elder migrants and their families including health care support; Neighborhood-based work and civic engagement with and for migrants; Promotion of cultural participation and intercultural dialogue as components of the strategic concept "interculture"; Reduction of inequalities and discrimination; Controlling and evaluation	– In same field of intervention: Organization of and participation in large religious events (Ramadan Market, Festival of Cultures, Action Week Colorful City of Hamm, Interreligious Peace Prayer) as indicators for successful intercultural and interreligious dialogue as well as identification of Muslims with their city

In contrast to former compensatory integration approaches, shortfalls of the "receiving society" shall now also be attended to. This is manifested in the existence of the intervention field "intercultural opening/orientation of the administration" in the integration plans of both cities. It is also reflected in the effort to create a "welcome culture" as opposed to a "culture of barriers" as signaled by the same two administration officials in the city of Hamm:

> . . . for these people [labor migrants] the experience during two or three generations was better not to interact with the city administration. The attitude was like "I go there, get my extension [of residence] and that was it." We have been trying for the last twenty years . . . to do the opposite of what we used to do and to say: "We are service-oriented. We . . . want to help you; we as the administration are here for you. We are not here to prevent, we are here to make things work." (H01S)

Neither in Hamm nor in Duisburg do "religion or interreligious dialogue/activities" constitute fields of intervention in their own right. But in Hamm, the cooperation with religious actors, such as mosque communities, is mentioned as one component of the intervention field "neighborhood-based engagement with and for migrants."

The organization of and participation in interreligious activities amongst other urban events is pictured as an indicator for successful intercultural and interreligious dialogue, as well as for the identification of Muslims with their city—corroborating the tendency stated in the literature (Tezcan 2006; Nagel 2012) to treat interreligious activities as a locus of intercultural understanding.

The equating of culture/interculturality with religion/interreligious events is, however, not occurring to the same extent in the city of Duisburg. Its integration concept expresses distance towards religion instead, pointing out that cultural activities in the city shall be "guided by a secular culture for all activities, including religious ones" (Integrationskonzept Duisburg: 20, translated quotation). However, religion constitutes one of eight focal fields of intervention of the Duisburg Council of Integration, which forms the local representative body of foreigners (Duisburg integration concept: 17). Hinting at legal obligations to attend to principles of religious neutrality, skepticism with regard to handling religion in urban policies is also voiced by an official of the Duisburg integration office:

> [We] as the city do not see ourselves as initiators for urban-wide interreligious encounters. This is simply not possible, because of the separation between State and Church, which we also need to attend to at the municipal level. This means that, of course, if religious communities invite us we do try to attend an event always depending on resources, time, and interest. (D07S)

In the perception of the city's religious communities, this skepticism, coupled with a certain indifference towards religion, may lead to interreligious dialogues becoming co-opted by the integration discourse. As one representative of the Jewish community in the city states their concern:

> . . . especially the welfare associations ("Wohlfahrtsverbände") struggle with interreligious topics, so intercultural issues are treated instead. The [Workers' Welfare Organization] . . .[8] in particular, their representatives say: "We do not have anything to do with religion!" . . . We used to do a celebration of religions ("Feier der Religionen"), which I found very nice and important. However, in the end, events that include a religious message have faded away somehow. (D10R)

But independent of the respective attitudes towards religion in the two cities, the relatively recent attention towards migrant religions and particularly Christian-Muslim dialogue has been without a doubt stimulated by Islamic terrorism (also view Nagel and Kalender 2014; Klinkhammer et al. 2011; Dinham et al. 2009: 7). It is largely against this background that interreligious and intercultural dialogue are framed as instruments for getting to know the "other," preventing intercultural conflicts and fostering (global and local) understanding and cohesion (Kufen 2008: 13).

Most of the interviewed state and religious representatives indicate that global or sometimes local incidents (for example, the prominence of the far-right movement in Hamm and its public demonstrations, and also Islamic/Salafist extremism) constitute sources of the political engagement with interreligious initiatives. City officials and religious representatives, the latter particularly from Muslim faith communities, in both cities highlight that interreligious dialogue has made an important contribution to lowering potential or actual tensions related to religious themes: "One has come to know each other" (D07S). Hence, in the following section we will turn our focus to the actual patterns of participation of public authorities in interreligious activities.

Making things work: interreligious activities and urban participation on the ground

Indeed, how public stakeholders conceive religion and integration also shapes their understanding of and their involvement in interreligious activities. In this section, we explore the interplay of state and societal actors in interreligious activities on the ground as a particular subfield of the urban governance of religious diversity. After describing the interreligious "landscape" of both city cases, we will analyze the modalities and degrees of public-private cooperation in these activities comparatively, concluding with a preliminary typology of state involvement in the interreligious field.

Interreligious activities and state involvement in Hamm

Even though the city of Hamm is considerably smaller than Duisburg, it exhibits a greater number and variety of interreligious activities, such as peace prayers, discussions groups, dialogue meetings, intercultural festivals, and concerts, a guided interfaith tour, and an interreligious installation in a public park.

Among these activities the interreligious "peace prayer" appears to be most established as it looks back on a history of more than a decade. It brings together representatives from different religious traditions, such as Sunni Muslims, Alevites, Jews, Bahá'í, Hindus, Orthodox Christians, Catholics and Protestants, and takes place in a church. Liturgically, it is based on a series of readings or recitations on the significance of peace within each religious tradition followed by a common song ("We shall overcome") and a joint prayer for peace. The municipal department of social integration takes part in the preparation meetings, but does not play an active role in the interreligious performance itself.

Another important format are discussion groups ("Gesprächskreise"). In contrast to peace prayers, these groups occur on the level of neighborhoods rather than on a city-wide level. As a consequence, they reflect the religious composition of their respective localities involving representatives from several Christian and Muslim denominations. Since they exhibit a much higher degree of *intra*religious diversity than other interreligious activities, they tend to create a forum for marginal groups, such as evangelical or Muslim minorities. Given their local scope, discussion groups seek to improve religious conviviality through interreligious understanding about practical matters of everyday life (rather than abstract theological matters).

Apart from the annual peace prayer and local discussion groups, the city of Hamm has come up with other, more unusual, interreligious formats. A good example is a multi-religious installation entitled "Gates of World Religions" ("Tore der Weltreligionen"), which was publicly inaugurated in 2012 in the Lippe Park, a former mining area that has been extensively re-landscaped. The installation consists of five huge copper gates, which represent Christianity, Judaism, Islam, Buddhism and Hinduism as "World Religions."[9] The gates are supposed to form a space of interreligious encounter and to provoke thoughts on religion and religious diversity. The original proposal was put forward by the Christian-Muslim discussion group in Herringen-Pelkum and further developed by the department of urban planning. A planning officer described his role in the process as follows:

> When this place was planned, there were fierce discussions . . . : How many gates are there? In which order should they be arranged? I can moderate such a process and take an external perspective, since it was not my idea and I have some emotional distance . . . Religion is always an emotional topic and that's all right, and thus I think that public administration has a good chance to take a neutral stance. (H08S).

The quotation points to the challenges of planning a multi-religious site that does justice to all religious traditions involved. The material incarnation of religion in public art poses questions as to basic concepts of salvation, which find expression in the gate metaphor. Further, it entails the need to make the relationships between religious communities explicit in a given spatial arrangement and give it a material form. At the same time, the city official introduces a distinction between "emotional" religion and neutral administration in order to claim the role of moderator.

The utilization of interreligious discussion groups as political opportunity structures was taken one step further in 2010 when the Integration Office initiated a new group in Bockum-Hövel, a neighborhood where one out of three people has a migration history. In our interviews, urban authorities tended to downplay their own role in this process: "It is important for us that we only initiate ("anstoßen") and attend ("begleiten") these discussion groups. We are not the organizers, but want the religious communities to organize themselves" (H01S).

At the same time, the added value of the intervention is emphasized: "Unlike the churches we are in touch with many different religious groups and therefore act as initiators and moderators" (H01S). Since the launch of the discussion group, the Integration Office has also provided translation services and supported the group in acquiring public funding for its activities. While the Integration Office seems to understand its participation as catalytic, the effects on the visibility and agenda of the discussion group are considerable. In contrast to older initiatives in Hamm's neighborhoods of Herringen-Pelkum and Wiescherhöfen, with a history of more than twenty years, the new group is the only one to maintain a website to communicate their work to a broader public. According to the website, the main objective of the initiative is "to get to know each other better, to improve conviviality in the neighborhood and to conduct joint projects," a goal which resonates well with the above-mentioned integration measures.[10]

To which extent do interreligious activities in Hamm stimulate what might be called a spiritualization of the city? Despite their innovative character, traditional religious institutions and authorities continue to play a crucial role in all types of interreligious engagement; hence, interfaith settings do not automatically go along with de-institutionalization. At the same time, all formats exhibit a sense of crossing or blurring denominational boundaries: local discussion groups seek to put forward an emphatic sense of neighborhood, based on an (alleged) interreligious consensus of values. Peace prayers offer a platform for multi-religious celebration and build strongly on what might be called a shared spiritual experience of ritual conviviality, while the "Gates of the World Religions" literally inscribe religious pluralism (albeit restricted to the so-called "Big Five") in a public space. This visible gesture of spiritualization is all the more powerful since the Lippe Park represents the overall transformation of the city from an industrial to a post-industrial metropolis.

Interreligious activism and state involvement in Duisburg

In contrast to Hamm, official interreligious activities seem to play a minor role in Duisburg. The city itself neither initiates nor organizes interreligious events (apart from occasional receptions administered by the mayor's office). This, however, is not to say that there are no interreligious activities in Duisburg. Instead, we see a number of local initiatives taking part in interreligious events on the ground, such as a Christian-Muslim dialogue circle, interreligious school services, and neighborhood festivals, as well as panel discussions on issues of social integration.

The official "interreligious reception" is held in the town hall and formally hosted by the mayor of Duisburg. In recent years, invitations have been extended to more than a hundred religious communities. According to a member of the mayor's office, the reception goes back to a Christian-Muslim round table in the town hall that took place about fifteen years previously. He indicates that the change in format and the extension of participation has significantly altered the character and impact of the event: "unfortunately there is something lacking, as we observe that structured dialogue, thematic dialogue, is replaced by loose networking, standing together and enjoying oneself. We increasingly miss a more substantial value ("Nährwert") as far as contents are concerned" (D11S).

As a consequence, the intention has been voiced to bring some more meaningful matters back into the event and to provide insights into the religious conduct of life ("Glaubensleben"), be it through thematic inputs or by performing traditional rituals. The desire to revitalize religious contents reflects an overall post-secular orientation in the mayor's office. This is closely associated with the person of the former mayor who is described as a devout, practicing Catholic with a particular openness towards Muslim groups: "During Ramadan he spent almost every night in a mosque community, being present and speaking in the name of the city" (D11S). As a matter of fact, this manifestation of personal religiosity among urban officials in Duisburg stands in remarkable contrast to the overarching secularist policy paradigm of the city, a tension discussed later in this chapter.

In addition to interreligious receptions in the town hall and Iftar invitations, urban stakeholders are involved in *interreligious meetings* that take place annually on the premises of different religious communities. Here, the mayor or a deputy gives a welcome address and actively takes part in the event. Moreover, we find a number of intercultural workgroups on the level of single administrative units, such as the department of youth welfare. At the same time, the advisory board of the Islamic Civic Center ("Begegnungsstätte") in Duisburg Marxloh has become an institutionalized contact zone between different religious and urban stakeholders. The board involves representatives from the neighborhood, political parties, churches, and the urban development agency. It has proved crucial for the smooth realization of the representative Merkez Mosque, which came to be known as the Marxloh Miracle ("Das Wunder von Marxloh").[11] To complement this, the city of Duisburg has launched a project to develop intercultural competencies within the public administration: trainees are supposed to undertake a two-week internship in a migrant organization, such as the Islamic Civic Center meeting center (D18R).

Moreover, the Marxloh Miracle offers a good opportunity to shed some light on the complex interreligious interaction between Muslim groups and the substantial Jewish community in Duisburg. The spokesman of the Jewish community attended the ground-breaking ceremony of the Marxloh mosque and remembered:

> I was allowed to speak at the ceremony. This is where I said that sentence which made the Muslims happy and Ralph Giordano [German atheist activist] angry: "we stand before you, by your side, and behind you." We have shown the Islamic Civic Center how to put together their community budget. They just did not know all this stuff. Hence, we have a very close relationship. (D10R)

This reflection indicates a central mechanism of boundary transgression between Jewish and Muslim groups, which is also characteristic for interreligious activities in general, such as building a religious phalanx vis-à-vis a secular or atheist counterpart, for example. At the same time, the emphatic assurance of support (most noteworthy in the words of an ancient Christian travel blessing) has a certain paternalistic appeal to it pointing to a fundamental asymmetry between the well-endowed and organized Jewish community and the enthusiastic yet chaotic Muslim groups.

Apart from the above-mentioned interreligious activities, orchestrated on the city level, and extensively supported by public administration and local policy makers, we have found a variety of interreligious events being organized on the grass-roots level of civil society with no or only indirect connections to public authorities. These bottom-up initiatives comprise a Christian-Muslim *discussion group*, which was started by Christian theologians and, unlike similar groups in Hamm, is explicitly dedicated to theological exchange, addressing topics such as creation, sin, or salvation. Moreover, schools appear to be an important hub of interreligious activities in Duisburg. Our respondents hinted at a number of *interreligious school services*, most of which had grown out of ecumenical celebrations and included an imam or hodscha, as well as a Catholic and a Protestant priest. Widespread occasions of these services include enrollment and graduation.

Other interreligious activities on the ground put a stronger emphasis on fostering a local sense of community across religious boundaries. These include neighborhood festivals, which are co-organized by local religious and cultural groups, as well as issue-specific round tables. Another branch of activities pays tribute to the industrial history of the region that was bound up with coal and steel production, for example, an annual festivity in honor of St. Barbara, protector of the miners. This takes place in an old mine in Duisburg-Walsum and came to include not only Catholic and Protestant dignitaries, but also an imam, thus reflecting the high proportion of Muslim workers in the mining industry. Likewise, a monthly "Political Night Prayer" ("Politisches Nachtgebet") has been conducted for almost ten years in several churches of the city, slowly shifting its focus from labor disputes to more general matters of equality and conviviality throughout the city (D14R). These prayers are conducted by labor unions, local Catholic and Protestant communities as well as social political Christian bodies, such as the pastoral service in the working environment ("Kirchlicher Dienst in der Arbeitswelt"). The close collaboration between Christian and social democratic associations on the ground is particularly noteworthy given the social democratic roots of Duisburg's political culture. These roots are often invoked in order to explain a general distance from religiosity in Duisburg's higher public administration.

Urban participation in interreligious initiatives: a preliminary typology

Apart from their descriptive value, the previous observations may serve as a starting point for a preliminary typology of urban participation in interreligious initiatives. Considering different degrees of intensity (active vs. passive) and quality (formal vs. substantial) we can distinguish between (at least) five types of public intervention: attendance, technical assistance, moderation, networking and initiation. Attendance refers to the presence of public authorities in interreligious activities (in contrast to more abstract forms of involvement, such as patronage or board membership). Even if they do not participate actively, the presence of a mayor or local police officer may change the character of interreligious exchange significantly. Technical assistance refers to a more active yet formal pattern of state support and can involve logistic and organizational assistance by public administration agencies as well as infrastructural or monetary assistance. Moderation, in contrast, refers to more substantial modes of participation including the facilitation of thematic discussions, translation, and mediation in the case of inter or intrareligious conflict. Finally, public authorities may actively administer interreligious activities through a targeted consultation and invitation of religious communities, which appear to be meaningful for the local agenda of integration politics (networking). Likewise, and more substantial in scope, they may even initiate interreligious activities on their own.

Comparative summary and perspectives

Our results illustrate the variety of interreligious activities in the two cities. They point to different settings and modes of cooperation between urban authorities and religious actors in the context of integration politics. These shall now be put into a comparative perspective, first with regards to the general orientation of integration policies, and second considering the patterns of public-private cooperation and their impact on the local religious field. Finally, perspectives for integration-related outcomes will be formulated.

Concerning the integration plans, the underlying goals are quite similar in the two cities in that emphasis is placed on aspects of intercultural understanding and dialogue. These are to be achieved through a combination of education and employment measures directed at immigrants and strategies of intercultural opening targeted at key actors and institutions of the receiving society. Moreover, the participation of civil society, such as religious or interreligious organizations, in the formulation and realization of integration plans is considered essential in both cities.

While interreligious encounters increasingly function as platforms for enhancing intercultural understanding and integration, the underlying political culture differs significantly in the two cities. While Hamm stands for a *post-secular setting*, which explicitly acknowledges religion and spirituality as integral parts of urban life-worlds, Duisburg represents a more restrictive notion of *state neutrality*, a position that also resonates in its integration concept. According to this view, too much public involvement with religious communities is to be prevented; instead, religion is allocated to the private sphere or to the grass-roots of civil society. However, the boundary between public neutrality and private religiosity is continuously blurred by urban authorities activating their own personal beliefs in order to get in touch with religious groups. At a first glance, the difference of political cultures seems to correspond to established factions of party politics with the Christian Democrat government of Hamm being friendlier towards religion than the Social Democrat government in Duisburg. At a second glance, however, we have found substantial collaboration between labor unions and religious actors in Duisburg, which points to a strong milieu of spiritual socialism beyond the official rhetoric of religious reluctance.

Partly related to the different political cultures of the two cities, urban authorities tend to take different roles within interreligious activities in both cities. In Hamm, representatives of the Integration Office have joined existing initiatives and come to play an active role as moderators or translators. Moreover, they have initiated a number of new interreligious activities, which either emulate established formats, such as local discussion groups, or constitute formats in their own right, such as the "Gates of World Religions." In Duisburg, public authorities provide technical and monetary assistance for a limited number of official interreligious events, but are not substantially involved. At the same time, some policy and administration actors invoke their personal religious affiliation and attitudes when dealing with religious communities, and thus introduce an informal level of interreligious encounter within the formal structure of their office.

Both city cases differ remarkably regarding the general locus of interreligious activism. Even though Hamm is much smaller, interreligious activities are organized on the neighborhood level. Duisburg, in contrast, stands for a more centralized approach, where public interreligious events are orchestrated through the mayor's office. At the same time, however, there are a number of interreligious grass-roots initiatives that take place, more or less unnoticed by urban authorities.

Regarding the repercussions of state involvement within the interreligious field, we find quite a number of religious communities participating in official interreligious activities some of which are more prominent than others. Central actors (in terms of influence, frequency and intensity of participation) include the Protestant and the Catholic parishes, and the well-organized Turkish Muslim association, the Turkish-Islamic Union for Religious Affairs (DITIB). Here, it is noteworthy that the Islamic Civic Center in Duisburg functions as a hub for Christian-Muslim dialogue and tends to diminish theological and organizational differences among Muslim communities. In contrast, interreligious activism in

Hamm reflects intra-Islamic diversity more explicitly, by systematically incorporating a Moroccan mosque community as local partner, for example. The periphery of religious participants is much more heterogeneous and includes smaller Christian denominations, such as orthodox or evangelical groups. In the case of Hamm, it is remarkable that the big representative Hindu temple plays a much more marginal role than its prominence in the official self-portrayal of the city would suggest. At the periphery we also find "new" religious communities, such as the Bahá'í in Hamm and the New Apostolic Church in Duisburg.

At the same time, public-private cooperation in the interreligious field can result in the *exclusion* of religious communities. In both cities, there were examples of self-exclusion, such as the Russian Baptist community in Hamm and an Alevi community in Duisburg, for example. Both groups have been repeatedly invited, but decided not to join official interreligious activities. An Alevi representative explained that the main reason for not joining was the uncritical stance of urban authorities towards "Islamist" groups, such as Millî Görüş (D22R). Likewise, the spokesman of the Jewish community in Duisburg pointed out that he would not officially cooperate with Millî Görüş as they were "openly anti-Semitic" (D10R). Another, more prevalent, mechanism of exclusion is not to invite certain groups. In both cities, there were substantial communities of Mormons and Jehovah's Witnesses, who have not been actively addressed. In Duisburg, there is also an active network of urban witches and practitioners of neo-pagan rituals, based at a shop called Eldemalu, who did not take part in any interreligious activity at all. The same applies to the local Buddhist and Bahá'í communities. It is difficult to assess empirically whether a given group had not been invited or refused the invitation for one reason or other. However, the restriction of interreligious activities to the Christian-Muslim encounter reflects the previously mentioned integration-security nexus in local policies. This and the focus on a set of so-called "world religions" (Christianity, Islam, Judaism, Hinduism, Buddhism) also resonates well with earlier observations of the field (Nagel 2012: 255–6).

Moreover, our results indicate that the emphasis on Christian-Muslim dialogue is strong even if there are other prominent communities from the world-religious spectrum, such as the Hindu community in Hamm or the Jewish community in Duisburg. A potential explanation might be that neither Hindus nor Jews have become subject to the above-mentioned security discourse. In addition, the Jewish community in Duisburg is embedded in other, highly institutionalized networks of public support (through a treaty with the federal state of North Rhine-Westphalia, for example) and interreligious understanding (such as the Society of Christian-Jewish Cooperation).

Finally, we ask: what are the outcomes of increasingly cooperative forms of (inter)faith governance for urban integration and managing religious diversity? Our results suggest that new networks have effectively been established and given way to an increased acknowledgement of religious actors and concerns. Moreover, the incorporation of interreligious initiatives into urban political agendas to build local social capital and foster social cohesion may enhance the innovative capacity, which is characteristic of the interreligious movement as such. As interreligious activities involve a continuous exploration of the margins of religious traditions, they can become incubators of new and hybrid religious worldviews (such as dialogue meetings) or practices (such as peace prayers). Moreover, the institutionalization of these activities in the framework of urban integration politics may trigger the evolution of innovative forms and formats of interreligious expression and exchange such as the "Gates of the World Religions" or the guided interfaith tour in Hamm. However, this does not mean that cooperative forms of (inter)faith governance invariably lead to an adequate accommodation of religious diversity in the local context. Rather, our results indicate that state involvement is likely to reinforce existing mechanisms and tendencies of exclusion, if, for example, dialogue with smaller (and "problematic") communities is shunned for political reasons. Additionally, our findings suggest that this tendency is more likely to manifest in larger cities with a more centralized approach towards public interreligious events than in smaller and more decentralized urban settings. However, more comparative research is required to further probe into the key factors influencing integration-related outcomes.

Notes

1 The document was updated in 2010 and renamed "National Action Plan Integration."

2 Following Haynes and Hennig, the public sphere is located in the environments of civil society, political society, and the governmental arena. It is to be distinguished from both the private sphere of the individual or household, and the religious field (Haynes and Hennig 2011: 2).

3 However, according to Cloke and Beaumont (2013), the spatiality of post-secularity may well extend into other areas, such as wealthy suburbs, through personal flows related to charity action, for example (Cloke and Beaumont 2013: 45).

4 Eade (2011) indicates that what may correctly be labeled as a "secularist" urban policy and planning tradition, holds only true for a very limited period of time in a very specific geopolitical context: the one of the post-World War II European Welfare State (Eade 2011: 154).

5 The project was entitled "Interreligious Activities and Urban Governance in the Ruhr-Area" and funded by the Mercator Research Center Ruhr (MERCUR).

6 In this regard, it is also interesting that the previous integration plan of the city of Hamm of 2003 focused on language promotion as one principal sector. The subsequent one of 2008 involves a considerably larger set of activities.

7 This analysis of the integration plans of the two studied cities draws on the integration dimensions developed by the German migration sociologist Hartmut Esser (1980: 22f.): structural (professional position, economic capital), cognitive (language, knowledge about the host society), social (inter-ethnic contacts and relations, networks.) and identification (sense of belonging to the host society).

8 "Arbeiterwohlfahrt" or AWO in German.

9 This selection is particularly noteworthy as there are no substantial Jewish or Buddhist communities in Hamm. Thus, it points to a normative and exclusive notion of "world religion" in which many interreligious activities are implicitly rooted.

10 http://www.cig-bockum-hoevel.de/wer-wir-sind/index.php, last access: August 22, 2015.

11 http://www.spiegel.de/spiegel/print/d-50034732.html, last access: August 22, 2015.

References

Baker, Chris (2008), Seeking Hope in the Indifferent City – Faith-based Contributions to Spaces of Production and Meaning Making in the Postsecular City. Paper given by Chris Baker at the Association of American Geographers annual conference in Boston, MA, April 2008.

Baraulina, Tatjana (2007), Integration und interkulturelle Konzepte in Kommunen, *Aus Politik und Zeitgeschichte* 22–3: 26–2.

Beaumont, Justin and Chris Baker (2011), Postcolonialism and Religion: New Spaces of "Belonging and Becoming" in the Postsecular city, in Justin Beaumont and Chris Baker (eds), *Postsecular Cities*. London and New York: Continuum, 33–49.

Bommes, Michael (2007), Integration – gesellschaftliches Risiko und politisches Symbol, *Aus Politik und Zeitgeschichte* 22–3: 3–5.

Cloke, Paul and Justin Beaumont (2013), Geographies of Post-secular Rapprochement in the City, *Progress in Human Geography* 37(1): 27–51.

Dinham, Adam and Vivien Lowndes (2009), Religion, Resources and Representation: Three Narratives of Faith Engagement in British Urban Governance, *Urban Affairs Review*, 43(6): 817–45.

Dinham, Adam, Robert Furbey and Vivien Lowndes (2009), Faith in the Public Realm, in Adam Dinham, Robert Furbey and Vivien Lowndes (eds), *Faith in the Public Realm. Controversies, Policies and Practices*. Bristol: The Policy Press, 1–18.

Eade, John (2011), From Race to Religion: Multiculturalism and Contested Urban Space, in Justin Beaumont and Chris Baker (eds), *Postsecular Cities*. London and New York: Continuum, 154–67.

Esser, Hartmut (1980), *Aspekte der Wanderungssoziologie. Assimilation und Integration von Wanderern, ethnischen Gruppen und Minderheiten. Eine handlungstheoretische Analyse*. Neuwied and Darmstadt: Luchterhand.

Gale, Richard (2008), Locating Religion in Urban Planning: Beyond "Race" and "Ethnicity"? *Planning, Practice and Research*, 23(1): 19–39.

Griera, Maria del Mar and Maria Forteza (2011), New Actors in the Governance of Religious Diversity in European Cities, in Jeff Haynes and Anja Henning (eds), *Religious Actors in the Public Spheres*. London and New York: Routledge, 113–31.

Haynes, Jeff and Anja Hennig (2011), Introduction, in Jeff Haynes and Anja Hennig (eds), *Religious Actors in the Public Sphere*. London: Routledge, 1–13.

Healey, Patsy (2007), *Urban Complexity and Spatial Strategies*. London and New York: Routledge.

Klinkhammer, Gritt, Hans-Ludwig Frese, Ayla Satilmis and Tina Seibert (2011), *Interreligiöse und interkulturelle Dialoge mit MuslimInnen in Deutschland. Eine quantitative und qualitative Studie*. Bremen: Institut für Religionswissenschaft und Religionspädagogik der Universität Bremen.

Krech, Volkhard (2008), Bewegungen im religiösen Feld: Das Beispiel Nordrhein-Westfalens, in Markus Hero, Volkhard Krech and Helmut Zander (eds), *Religiöse Vielfalt in Nordrhein-Westfalen. Empirische Befunde und Perspektiven der Globalisierung vor Ort*. Paderborn: Schöningh, 24–43.

Kufen, Thomas (2008), Geleitwort, in Markus Hero, Volkhard Krech, and Helmut Zander (eds), *Religiöse Vielfalt in Nordrhein-Westfalen. Empirische Befunde und Perspektiven der Globalisierung vor Ort*. Paderborn: Schöningh, 13–14.

Leggewie, Claus (2009), Religion als Hemmnis und Medium lokaler Integration, in Frank Gesemann and Roland Roth (eds), *Lokale Integrationspolitik in der Einwanderungsgesellschaft*. Wiesbaden: VS Verlag für Sozialwissenschaften, 593–607.

Mayntz, Renate and Fritz W. Scharpf (1995), Steuerung und Selbstorganisation in staatsnahen Sektoren, in Renate Mayntz and Fritz W. Scharpf (eds), *Gesellschaftliche Selbstregelung und politische Steuerung*. Frankfurt am Main: Campus, 9–38.

metroZones (2011), Urban Prayers, in metroZones (eds), *Urban Prayers. Neue religiöse Bewegungen in der globalen Stadt*. Hamburg und Berlin: Assoziation A, 7–24.

Nagel, Alexander-Kenneth (2012), Vernetzte Vielfalt: Religionskontakt in interreligiösen Aktivitäten, in Alexander-Kenneth Nagel (ed.), *Diesseits der Parallelgesellschaft. Neuere Studien zu religiösen Migrantengemeinden in Deutschland*. Bielefeld: Transcript, 241–68.

Nagel, Alexander-Kenneth and Mehmet Kalender (2014), The Many Faces of Dialogue. Driving Forces for Participating in Interreligious Activities, in Katajun Amirpur, Wolfram Weisse, Anna Körs and Dörthe Vieregge (eds), *Religions and Dialogue. International Approaches*. Münster: Waxmann, 85–98.

Penta, Leo and Werner Schiffauer (2011), Nur in der Stadt kannst Du rein religiös sein. Ein Gespräch über Politik und Praktiken religiöser Gemeinschaften in Berlin, in metroZones (eds), *Urban Prayers. Neue religiöse Bewegungen in der globalen Stadt*. Hamburg und Berlin: Assoziation A, 249–70.

Sandercock, Leonie (1998), *Towards Cosmopolis. Planning for Multicultural Cities*. New York: Wiley.

Sandercock, Leonie (2003), *Cosmopolis II. Mongrel Cities in the 21st Century*. London and New York: Continuum.

Sandercock, Leonie and Maged Senbel (2011), Spirituality, Urban Life and the Urban Professions, in Justin Beaumont and Chris Baker (eds), *Postsecular Cities*. London and New York: Continuum, 87–103.

Tezcan, Levent (2006), Interreligiöser Dialog und politische Religionen. *Aus Politik und Zeitgeschichte* 28–9: 26–32.

Weller, Paul (2009), How Participation Changes Things: “Interfaith”, “Multi-faith” and a New Public Imaginary, in Adam Dinham, Robert Furbey and Vivien Lowndes (eds), *Faith in the Public Realm. Controversies, Policies and Practices*. Bristol: The Policy Press, 63–81.

3 Preserving Catholic space and place in "The Rome of the West"

Tricia C. Bruce

Stainless steel and higher than any other monument in the United States, the Gateway Arch towers over the city of St. Louis as a symbol of westward expansion. Captured in its shadow along the Mississippi River's west bank is the Catholic Cathedral Basilica of Saint Louis IX, King of France—"The Old Cathedral." Its own impressive stature long precedes that of the proximate Arch. Founded in 1770, the parish served the entire population of Catholics in St. Louis, then arriving primarily from France. Fourteen large windows and a 40-foot nave illuminated worship space for the 15,000 parishioners it once claimed as the only Catholic church in the city. The "mother church" was the first cathedral west of the Mississippi, and the only building on the river's edge to remain when the derelict waterfront was cleared in 1963.

The Old Cathedral's transcendent presence marked the city of St. Louis as Catholic. But more than two centuries of change left the parish more cavernous than lively, more a relic than a core of Catholic activity in the city. Catholics' exodus from urban St. Louis in the wake of late twentieth-century suburban expansion decimated the Old Cathedral's parishioner base. Its registered parishioners constituted but a fraction of the church's seating capacity. The number of annual baptisms were a shadow of historic highs. Dwindling enrollment closed the parochial school long ago, and divisions of the parish's original geographic territory spawned numerous congregations competing for far fewer St. Louis Catholics.

What preserves the Old Cathedral, along with other historic landmarks of Catholicism in the city of St. Louis? And how does the new urbanesque—marked by religious adherents' wider geographic dispersion and increasingly diverse needs—mandate innovation to preserve religion's visibility?

This chapter focuses on the Catholic Church's entrepreneurial response to demographic change that dramatically altered Catholic presence in downtown St. Louis, Missouri. As the number and density of Catholic adherents declined and institutional resources shrank, St. Louis's Catholic leaders faced a simple question: how to save their beautiful, historic, downtown churches? The answer came in the form of "personal parishes" serving Catholics through common *purpose* rather than common geography.[1]

Building the Rome of the West: a history

St. Louis epitomizes Catholicism's immigrant emergence and rise in America. At the time of its 1826 inception, the St. Louis diocese was the largest in the United States. Successive waves of European immigration, new religious orders, a Jesuit college, and steadfast diocesan leadership elevated the archdiocese to a central place in American Catholicism. The city acquired the moniker "Rome of the West" with the first cathedral west of the Mississippi River and another cathedral that Pope Paul VI would later call "the most beautiful church in the new world." Hundreds of neighborhood parishes anchored Catholics to their churches and their churches to the city ("History" 2001).

But major changes after 1950 reoriented St. Louis Catholics' space, place and identity in the city. Highway development, mass automobile production, housing construction and modern shopping centers blurred the boundary between the city and its suburbs, fostering an urbanesque habitat. St. Louis's urban core shrank as its suburban edges bloomed. Catholics found new homes, neighborhoods, parishes and schools in the suburbs, while the downtown parishes emptied.

Intensified racial-ethnic relations and class divisions further bifurcated the city. Racialized zoning left St. Louis' history "bound up in a tangle of local, state, and federal policies that explicitly and decisively sorted the City's growing population by race" (Gordan 2008: 69). St. Louis's Cardinal Archbishop Joseph E. Ritter (from 1946 to 1967) ushered in new policies of racial integration for Catholic parishes and schools (Dolan 1985). But by 1970, some 60 percent of whites had migrated out of St. Louis's downtown (Gordon 2008). An increasingly African American population inhabiting downtown was more likely to affiliate with Protestantism than Catholicism.

The latter half of the twentieth century, then, marked the exodus and decline of Catholic presence in the city of St. Louis proper. As one downtown pastor stated during our interview, "The neighborhood has changed quite a bit."[2] Fifty-four Catholic parishes in the Archdiocese of St. Louis closed between 1950 and 2000. Historically, European immigrant-serving parishes, in particular, felt the pain of this neighborhood change. French, Hungarian and Bohemian parishes, by and large, no longer served French, Hungarian, or Bohemian Catholics. A former St. Louis bishop recalls:

> You have a church that might have been a Lithuanian church, and now there are no more Lithuanians coming. We might have African American and other people that come, who are welcomed there, because the church welcomes everybody to a Catholic church. But little by little, the Lithuanian character is no longer pressing and then boom: all of a sudden, it happens that we just don't have the people. We can't support the parish anymore.

The parish of St. Mary of Victories, for example, epitomized the changing Church in a changed city. Established before the Civil War, it had originally served

German Catholics amidst a commercial and residential hub of St. Louis. Later greeting post-World War II migrants, it was dubbed "the Hungarian church," and emerged as a nucleus of Hungarian cultural events. But the construction of St. Louis's Highway 55 in the 1960s—taking an upwardly mobile generation of mostly white Catholics out to the suburbs—left the church quite literally beneath the underpass, amidst aged, downtrodden warehouses and streets. Said a St. Louis pastor of the parish's fate:

> Some of these ethnic groups have either dissipated or assimilated into the broader culture, or they just simply moved out of the area that's nearby. Sometimes it was because of things outside their control, like St. Mary of Victories, which you see as you're going down 55. There was a flourishing parish of traditional Hungarians; it was a Hungarian neighborhood. But then the city planners decided to put a freeway through there.

Highway construction razed parishioners' homes, and suburban expansion eclipsed downtown. Downtown parishes left behind failed to successfully integrate their new neighbors. St. Louis's landmark Catholic churches stood as beautiful, historic, culturally rich—and largely empty—buildings.

Restructuring Catholic St. Louis today

The legacy of St. Louis's founding anchored the Catholic Church in a visible, physical and literal sense to urban St. Louis. Steeples filled the skyline. Churches stood mere blocks apart. Though scores of Catholics had moved, their church buildings remained. On Sundays, two out of three pews were empty. The Catholic population had fallen by half across areas of the city, but roughly the same number of churches remained open. As of 2004, 213 parishes served Catholics in the Archdiocese of St. Louis—a number that paralleled that of dioceses three times larger. With some 2,600 Catholics-per-parish compared to above 5,000-per-parish elsewhere (*Annuario* 2005), there were simply too many Catholic parishes serving too few Catholics.

That archdiocesan restructuring would emerge in this climate was not altogether surprising. Fewer Catholics, financial constraints and a shrinking number of priests made it impossible to maintain all the downtown churches. It was a situation common among major metropolitan areas undergoing reorganization (Weldon 2004). Financial hardship in the US Church intensified following revelations of sexual abuse that peaked in 2002, reverberating throughout the country (Bruce 2011). Like other dioceses experiencing resource contraction, the Archdiocese of St. Louis needed to reduce its number of parishes.

Despite engaging in important urban ministry efforts, many parishes simply did not have a stable parishioner base and sustainable funding. Downtown pastors were aware of this grim outlook. One pastor described seeing the writing on the wall, presuming his parish's inevitable closure during archdiocesan restructuring:

> In an immigrant population, they built parish churches that could be a cathedral anywhere in the newer dioceses. These big, dramatic structures. And now, unfortunately, the neighborhood has evolved and changed and it's not as Catholic, and the buildings need a lot of upkeep.

Under the leadership of Archbishop Raymond Leo Burke, the Archdiocese of St. Louis closed 31 city parishes in 2005.[3]

In closing parishes—which in practice meant reallocating the parish's territory and assigning parishioners elsewhere—the archdiocese also faced selling, razing, or repurposing the effected church buildings. St. Louis had an excess of beautiful churches where the number of Catholics no longer justified their territorial need. But the Catholic Church could not abandon its cathedrals. St. Louis's declining churches held a physical and symbolic presence in the city that was historically rich, magnificently built and culturally circumscribed.

Interviewees recall the dismay and resistance that this prospect evoked. "There's no way you can tear down a church like this," recounts a young St. Louis priest from the conversation that transpired in his church. The archbishop himself acknowledged the particular challenge of "the number of churches which are treasures either by reason of their architectural and decorative beauty or their history" (Burke 2005b: 3).

A lay diocesan staffer who participated in restructuring conversations described the situation as such:

> What happened in 2005, as we were contemplating closing a number of parishes, was that we had a number of very beautiful, historic church buildings that we were wishing that in a fantasy world you could just pick them up and move them to where the people were. But you couldn't do that, so you had to figure out what to do with these buildings. Yes, many of them were on historic registers, either national or local historic registers, but we have torn down historic buildings before, historic churches. That wasn't totally the issue, but I guess as sensitivities toward retaining historic buildings became more prominent in the last two or three decades than it was maybe back in the '40s and '50s, we had to figure out how you could balance having territorial parishes and then what to do with these really unique buildings.

How was the archdiocese to save its majestic, historic, urban religious landmarks? What the archdiocese needed was a reason to keep the churches, a reason that was not reliant upon the relative density of the Catholic population. As a former bishop put it, "A beautiful building that a bishop thinks should be preserved – obviously he's going to think of different ways to preserve it, without doing something artificial." He adds, "The saving of a building is not a justifiable means for inventing something that is not true. But it could be a providential occasion."

And so, the Archdiocese of St. Louis resurrected the notion of the "personal parish." Formerly, "national parishes" had deeply inscribed culture into St. Louis's parish past. Personal parishes offered a contemporary, formal means of inscribing culture into St. Louis's parish present. Landmark Catholic buildings could be saved, accordingly.

Personal parishes

Unlike their more ubiquitous territorial counterparts, personal parishes need no defined geographic territory within a diocese. Rather, they are designated to meet a special purpose, in service to an identified need or population of Catholics. Catholic Canon Law outlines their use as such:

> As a general rule a parish is to be territorial, that is, one which includes all the Christian faithful of a certain territory. When it is expedient, however, personal parishes are to be established determined by reason of the rite, language, or nationality of the Christian faithful of some territory, or even for some other reason. (*Codex* 1983: Canon 518)

Like their erstwhile national parish iterations, personal parishes may dedicate service to a specific ethno-cultural or linguistic community. But personal parish designation need not be ethnicity- or language-based. Church law further empowers bishops to identify other, non-ethnic-based needs or purposes best met via non-territorial (personal) parishes.

Catholicism's more typical reliance upon geographic parishes suggests that the Church is comparatively more "rooted" in urban environments (Gamm 1999) than other denominations.[4] Personal parishes both affirm and challenge this idea of rootedness, as applied in the suburban and the urbanesque. They draw not just from immediate geographic surroundings, but from anywhere in the greater vicinity. This model of parish building can work especially well in metropolitan contexts, where voluntary participation can lead to new styles of belonging (Wedam 2000).

Personal parishes introduce forms of religious community less reliant upon proximity. They embody the idea of "community without propinquity"—eschewing place for purpose—as per the title of a 1963 paper by urban theorist Melvin Webber. Though personal parishes also evoke the notion of the "de facto congregation" (Warner 1994) in relying upon voluntary membership, they are distinctive because they also require formal establishment and sustained ties to the diocese. Personal parishes find their livelihood through identified needs arising from changed religious ecologies. Pope John Paul II spoke of the unique challenges presented by urban settings, in particular: "Because of the particular problems they present, special attention needs to be given to parishes in large urban areas, where the difficulties are such that normal parish structures are inadequate" (John Paul II 1999: no. 140).

Personal parishes gave the Archbishop of St. Louis a lesser-known but still institutionally sanctioned way to keep church buildings open when the Catholic population did not warrant their salvation. Though not wholly new, given the historic presence of national parishes, their application in St. Louis presented a novel innovation in the interest of saving landmark buildings. Churches otherwise slated to close would become personal parishes. Summarizes one middle-aged man, a lay participant in the restructuring process:

> We solved [the question of saving historic buildings] by trying to figure out how we could manage territorial parishes—seeing that there were not only unique needs for national ethnic-based parishes, but also there were some parishes that we were able to keep and we found a reason to keep them.

Empty churches provided a solution looking for a problem. In the absence of "neighborhood" need, the Catholic Church in St. Louis had to strategically name other needs currently unmet by territorial parishes. If unmet needs could be identified, personal-parish status could be justified to meet them. Closed parishes could deliver both the impetus and the solution for restructuring.

Personal-parish designations would inject new wine into old St. Louis church wineskins, preserving the visibility and cultural presence of Catholicism in urban St. Louis. Recognizing this, one young St. Louis pastor describes:

> In the United States, we have to rethink so many different ways that we have been doing the church, doing the parish model. Because otherwise, we end up with these big husks of churches all along the city that once had been thriving and are now empty.

A bishop formerly in St. Louis echoes this sentiment, "The Church is very much alive and vibrant, so it moves and it lives, and changes, and evolves. Churches don't lead people to where they live—churches follow."

Innovating along these lines, restructuring parishes in St. Louis took the form of identifying niche cultural needs in the diocese, establishing personal parishes devoted to meeting those needs, and housing personal parishes in the historic church buildings of Catholic St. Louis that would otherwise close. This would justifiably and canonically preserve the city's Catholic spaces. It was cognizant innovation on the part of the archdiocese to maintain and newly conceptualize a Catholic presence in the city.

Needs in the "new urbanesque"

The modern urban agglomeration of St. Louis offered Catholic archdiocesan leaders an abundance of rationales to justify personal-parish status. Residential and commercial activities now spread beyond the urban core. Residents more readily traveled from home to work, to church, and to leisure activities. Visitors and travelers itinerantly interfaced with downtown sites. Class inequality undergirded persistent social need. An ever-more-cosmopolitan population introduced new forms of linguistic and cultural diversity.

Within each of these characteristics of the changed urban context, the Catholic Church in St. Louis found opportunity to identify unmet needs worthy of personal-parish status. Needs fell generally into four categories, which I label: (1) ethnic ministry; (2) distinctive liturgical forms; (3) social justice, and (4) spiritual tourism. Each provided an avenue through which to constitute new personal parishes—none reliant upon geographically proximate parishioners—to be housed in existing church buildings. All were established via official pronouncements from then-Archbishop Burke in February 2005.

Ethnic ministry

A handful of niche St. Louis Catholic communities readily presented the most transparent arena of need: distinctive ethnic and linguistic expressions of Catholicism. Half of the personal parishes named in 2005 served needs specific to ethnicity, culture, nationality and/or language. Stated one Latino pastor of this strategy, "It's a natural thing, especially in St. Louis: there are so many national parishes in its history."

A growing population of Asian and Latino Catholics in the city, for example, brought new languages, celebrations, practices and desire for cultural preservation. Largely itinerant and tangential in their inclusion at territorial parishes, first-generation migrant Catholics had experienced varying degrees of welcome and parish ownership during their St. Louis tenure. Few priests had the language capacity to serve non-English-speaking minority communities. Reconstruction provided an opportunity for the archdiocese to strategically consolidate newer immigrant subcultures into a handful of personal parishes. The communities could be given personal-parish status, and then moved into existing church buildings that had previously housed neighborhood Catholics.

St. Thomas of Aquin Parish, for example, became "a personal parish for the faithful of Vietnamese language and heritage," after being otherwise emptied through its closure as an English-speaking territorial parish. Unlike other new ethnic groups in St. Louis, Vietnamese Catholics had actually secured personal-parish status some ten years prior, but in restructuring were asked to relocate. Their original space was smaller and in decline. Restructuring offered new spatial options (that is, newly emptied churches). Thus, the first Vietnamese parish "home" was officially closed in 2005 and a new (old) home erected. A diocesan pastoral planner describes:

> [The new parish] was in an area where we had to consolidate three parishes into one. There just weren't enough Catholics to have three parishes. The Vietnamese people lived already in this area. The church itself happens to be some kind of a 1950s neo-something architecture. It's got a unique architecture. That just kind of worked out.

Vietnamese Catholics enlivened otherwise aged grounds in South St. Louis. Though English words and statues with European facades still characterize the church's interior, its occupants are now almost entirely Vietnamese, most born in Vietnam. Our Lady of Lavang now stands in an outdoor courtyard of the new personal parish, flanked by flowers and the prayers of a new crowd of faithful.

Personal-parish status married extant need and community with physical space and a renewed sense of welcome and pride for St. Louis' Vietnamese Catholics. A former bishop shared how this reciprocal exchange met simultaneous diocesan needs:

> It was obvious that the establishment of the parish for the Vietnamese corresponded to a real need. And it was a great blessing besides that, because of the people—it was the best of enculturation in the church. They were so happy to bring their culture into the church, and they were so happy to be able to, as Catholics, actuate their cultural traditions. So it was a beautiful wedding, if you will.

Those in the parish describe the union in similar terms, as one man, a parish leader, explains:

> The archdiocese knew our needs, but hadn't had the opportunity yet, until this was about to close. Then, you know, two things happened at the same time. They see the need, and we also expressed this to them. And then they saw the opportunity open.

As he conveys, the personal parish met Vietnamese Catholics' distinctive needs by granting them a space of their own to serve their vibrant community. "They feel more comfortable worshipping in their own language," explains the pastor. The restructuring process gave St. Louis' Vietnamese Catholics the option to move into an existing church building. The community had a home, and a church in the city could remain open.

Another new personal parish for Latino Catholics bridged cultural and language needs with the desire to save an impressive, cathedral-like parish church. St. Cecilia's was not an obvious choice for a Spanish-speaking ministry home; a fairly small group of Latino Catholics belonged to the previously declining parish community, then facing closure. But with few Spanish-speaking priests and a dispersed population of Latino Catholics, the archdiocese opted to maximize efficiency by consolidating the Spanish-speaking ministry into two personal parishes. "It's practical to do that in a place like St. Louis," shared a St. Louis pastor. The original pronouncement for St. Cecilia's read:

> St. Cecilia Parish is to become a personal parish for the faithful of Hispanic language and heritage. Other faithful in the area may also be members of the parish, engaging fully in the life of the parish and giving particular support to the service of our Hispanic brothers and sisters in the deanery and the archdiocese. St. Cecilia Parish is, moreover, to become the center for the Church's Hispanic Ministry in the city of St. Louis. (Burke 2005b: 14)

Described in the archdiocesan paper at the time, "It was determined that having a territorial parish and a personal parish for the Hispanic/Latino community in the same parish would provide enough membership and finances to ensure its long-term viability." New parishioners—identified and relocated—would allow the church to subsist. Another parish in the north of the city likewise concentrated formal service to "the faithful of Hispanic origin or language" (Burke 2005a).

Prioritizing efficiency, the combination of disparate Latino communities and necessary displacement of former, non-Latino parishioners did make some St. Louis Catholics angry ("That wasn't as smooth as everybody would have liked" suggests one lay participant in the process). It also arguably circumvented other territorial parishes' responsibility to minister to a growing Latino Catholic population. But St. Cecilia's was grand, historic—and now brimming with Latino faithful.

Polish Catholics, too, were given personal-parish status at St. Agatha's, a historic landmark church just blocks away from another historic landmark: the Anheuser-Busch brewery. The Polish designation was politicized in the wake of a highly publicized feud over ownership and finances at another historically Polish parish in the diocese, St. Stanislaus Kostka. That ordeal, begun in years preceding the 2005 restructuring, resulted in St. Stanislaus's pastor and board members being excommunicated. The church was no longer allowed to associate with the archdiocese. A newly designated personal parish at St. Agatha's, then, offered sanctuary to Catholic Poles desiring union with the archdiocese.

Finally, St. Joseph's durability was secured as "the personal parish for the faithful of Croatian language and heritage," with "special attention to the new immigration from Bosnia." A sanctuary of Croatian culture for more than a century, its renewed status as a personal parish ensured its livelihood for continued service to migrants, both old and new. Though the neighborhood looks far less Croatian today, generating a parish community from the wider city population allowed it to keep its purpose clear and its doors open.

Distinctive liturgical forms

Ethnic needs may have been the most expected based upon historical precedent, but they were not the only needs leveraged for personal parish status to save St. Louis churches. Catholic leaders expanded "ethnicity" into other forms of identity, culture and taste—bolstering while also broadening a legacy of ethnic accommodation.

Included in this category was the architecturally stunning St. Francis de Sales, newly designated as "an oratory for the faithful attached to the celebration of the Mass and other sacred rites, according to the liturgical books in force in 1962." The church would now exclusively offer the traditional Latin Mass—common across all Catholic Masses prior to the reforms of the Second Vatican Council of 1962–65.

St. Francis de Sales, with its German Gothic architecture and 300-foot spire, appropriately carries the moniker "Cathedral of South St. Louis." Holding a place on the National Registrar of Historic Places, its size, age and grandeur made it excessively expensive for the Archdiocese of St. Louis to maintain and thus a prime candidate for closure. Without its renewed status in 2005 as an oratory for the Traditional Latin Mass, its property and magnificent structure would have been sold and demolished.

Staving off this fate, Archbishop Burke reached out to the Institute of Christ the King, a society of priests with a special focus on the traditional ("extraordinary") form of the Mass. The institute carries a record of fundraising for restoring and revitalizing historic church buildings. The union was a "marriage made in heaven," as one participant in the process put it:

> We needed a place for the celebration of Latin Mass, and Cardinal [Archbishop] Burke was a large proponent of trying to bring in the Latin Mass in a more formal way than it was before. Previously, we had maybe one or two places where a priest would celebrate the Latin Mass and people would just go there. He wanted to create something more formal—like, here's an oratory that's specifically created for the celebration of the Latin Mass. He saw what a beautiful church St. Francis de Sales was. He really thought that would be a beautiful church for the Latin Mass. He brought in the Institute of Christ the King, and it was just a marriage made in heaven.

The church's traditional chair rails, protracted aisle, and majestic stature made it particularly well suited for the Latin Mass. Services now attract large families with young children, their mothers donning traditional lace veils. Silence shrouds the meticulous motions of priests and altar boys as they carry on the highly ritualized and long-standing form of Catholic liturgy.

"When [Cardinal Burke] created St. Francis de Sales," shared a priest whose own family belonged to the parish when it was still territorial, "he saved that church building. Because they were able to draw money from [the Institute] for people who are trying to invest in taking a beautiful work of art and trying to maintain it." Traditionalist Catholics now had a home in the city, and a beautiful one at that.

Yet another parish, mere blocks away from St. Francis, was formally closed as a territorial parish and resurrected as a "personal parish for the faithful attached to the spirituality of the Missionaries of the Holy Family, whose provinciate is located in the rectory of the parish." Their church met an identified need for ministry focused on the family, granting also a sustained home for the religious order of priests occupying it. The restructuring designation came as a welcome surprise. A man who is a leader in the parish recalls, "I don't think anybody even thought of the possibility of having a personal parish. All the parishes put together their reasons why they needed to stay open. Everybody's given that opportunity. And we did too, we were just too small." Though territory could not justify their salvation, special status as a personal parish did. The church remained open.

Social justice

A third category of need—rationalizing the protection of two more urban St. Louis churches—was that of social justice. Capitalizing upon a robust Catholic commitment to social justice, this category of personal parishes emerged from the disparity of resources and economic struggles of St. Louis's downtown residents.

At St. Cronan's, a sign hanging on the parish's exterior boasts: "It is good to be here . . . Celebrating 130 years at St. Cronan's." A Saturday morning visit encounters an impressive stockpile of clothing and household items, free for the taking. Low-income residents from the neighborhood walk away with garbage bags filled with needed items. Catholic Charities occupies what used to be the parish school, just across the street. Weekend Masses incorporate heavy lay-person involvement and an intentional welcome to "people of all ages and persuasions," as stated in the parish's mission, quickly apparent in the hodgepodge of attendees and prayers to accept all for who they are. A sign of peace offered midway through Mass virtually halts the service, as attendees walk freely around the sanctuary shaking hands and embracing. Songs evoke almost unison participation, even spirit-led movement.

"St. Cronan Parish is to be established as a personal parish dedicated to the apostolate of social justice," stated the pronouncement that secured the church's fate as a personal parish in 2005. Progressive and social justice-oriented Catholics had a home. The 130-year-old church would live on.

Some five miles closer to the Mississippi River, in the historic immigrant neighborhood of Soulard, stands another elaborate Gothic revival church with seating capacity for 3,000: Sts. Peter and Paul. As industry encroached, Catholics had fled to the suburbs. The old parish school had been converted into loft apartments. Today's smaller cadre of devoted attendees—nearly all of whom live outside the neighborhood—shrink in the opulence of the space. Here, too, in a context of homelessness and other forms of social need, the parish's ministry to the poor thrives. An annual "potty club" along the Mardi Gras parade route charges for clean restrooms as a fundraiser to alleviate neighborhood poverty. A homeless men's shelter operates in the church basement year-round; parishioners serve free meals daily. A full-size tour bus for the multi-parish performing group of "Young Catholic Musicians" resides in the parking lot.

"Sts. Peter and Paul Parish is to be a personal parish dedicated to the apostolate on behalf of the homeless and to the promotion of sacred music through the Young Catholic Musicians" reads the 2005 pronouncement. Though the modern Catholic population of Soulard could not support the density of territorial parishes in and near South St. Louis, the archdiocese nonetheless found a means of preserving Sts. Peter and Paul as a landmark and institution of social justice in the city.

To the question of why their parish was "saved," one parish leader in his mid-sixties expressed:

> That's a really good question, because we've been wondering that question a lot. We're a very vibrant, ministry-oriented parish. We have a reputation in

> the diocese as being very active. Even though we're small, we're mighty—if you would. And a very generous group of people, too.

Echoing their renewed vitality, the church now describes itself as "A Catholic parish in the historic Soulard neighborhood serving the poor."

Spiritual tourism

The fourth category of personal parishes engaged by the diocese—spiritual tourism—was perhaps the most unusual. There was no precedent historically, or even much of one nationally today, to designate personal parishes with a tourist-oriented purpose. But the most grand and highest-profile church in the archdiocese facing closure needed a way to stay open. And a creative personal parish status could offer that.

The Old Cathedral—neighbor to the Gateway Arch—simply could not close. It certainly no longer served residential Catholics in the way it once did. Downtown workers and visitors were now its primary, albeit occasional inhabitants. Territory could not keep it open, but purpose could. In restructuring, the St. Louis Catholic Archbishop found creative occasion to name the need: tourism. Outlined more specifically, the 2005 pronouncement declared:

> The Parish of the Basilica of St. Louis, King of France, founded in 1770 and the mother church not only of the Archdiocese of St. Louis but of a good part of the Midwest, is to remain a personal parish for those who have a special attachment to it, for reason of its history and sacred architecture. It serves as a spiritual oasis to the many visitors to the city for tourism or special events. It is also a place of devotion to the saintly Bishop Joseph Rosati, first bishop of St. Louis, whose mortal remains are interred in the basilica. (Burke 2005a)

Embodying a far different role than that of its origins in service to some 15,000 parishioners, the church would now take on an unparalleled role in representing Catholicism to the city and to the Midwestern United States.

Scaffolding and plastic now flank the cathedral's grand stone entrance, in the midst of renovations enabled through civic and church contributions. Stacks of replacement windows lean against work-trucks laden with paint, tarps and tools. The cavernous inner space smells at once old and new. A sacristan arrives midday via motorcycle in a black leather jacket to prepare the space for a thirty-minute weekday service whose sermon highlights unity amid difference. The Old St. Louis Cathedral remains open as a prime Catholic site in the city.

With needs and purposes identified that were not reliant upon territorial residency nor a neighborhood density of Catholics, the Archdiocese of St. Louis could justifiably establish personal parishes to meet those needs. The move was at once consistent and contradictory with their historical application of personal parishes as national parishes, in particular. A pastoral planner shares:

> Well, we always had personal parishes, so the concept of taking some of these territorial parishes and making them ethnic personal parishes . . . It didn't strike me as being terribly unusual. I think if you read the actual Canon Law associated with personal parishes, it talks about meeting certain people of nationalities, but also it talks about other unique [needs]. (Burke 2005b)

As to whether "need" is here being extended too far, eschewing the predominant model of territorial parish building (a "cop-out for not really seriously dealing with the demographics and the reality of what's happening," warns one St. Louis pastor), a bishop offers,"It all depends on the reason for the need. If it's something

Figure 3.1 Renovations underway at the Catholic Cathedral Basilica of Saint Louis IX, King of France ("The Old Cathedral"). Originally founded in 1770, it became a personal parish in 2005 "for reason of its history and sacred architecture," welcoming tourism and special events near the Mississippi River's west bank. Photo by Tricia C. Bruce.

to break away then that's not; that's not good. But if it's something to add to the beauty of a local church, I think it's important."

Preserving sacred space through "Community without Propinquity"

The 2005 establishment of personal parishes in St. Louis was not merely an organic outgrowth of extant ethnic enclaves. Each of the new (or old, as the case were) parishes did not mark a congregational community by neighborhood but by shared purpose, propinquity notwithstanding. Attendees now drive in to attend from around the diocese. The move showcases a key innovation in spiritualizing the new urbanesque: community is bound less by territorial proximity, instead more geographically dispersed—all over the agglomeration and beyond—and constituted from shared purpose.

Though each of the personal parishes was "new," none received a newly constructed physical space. All were embedded within existing churches—a core motivating factor for personal parish establishment in the first place. Even among those that remained in their same church structures, their former territorial parish designation succumbed to a personal-parish one to justify their continued existence. "Neighborhood" parishioners had to be reassigned to another geographic parish; the church structure then housed a personal parish community serving a specialized function. Catholics in St. Louis were quite literally picked up and moved into new churches. While church structures could not move, people could. Community is preserved; space is transformed.

Recalling the decision, a diocesan interviewee shared:

> To satisfy the needs of the diocese, we're going to have to close some parishes, and if we're going to keep a parish open, we may as well give them a place and it may as well be a good place. It's a beautiful place. So we got to keep it open while we close down some others.

In this way, each personal parish—as an interviewee put plainly—"saved another building." Combined with identifying niche purposes, personal-parish establishment redirected the fate of St. Louis's downtown Catholic spaces.

Urban religious restructuring in St. Louis was as much proactive as it was reactive: strategically and systematically reallocating sacred spaces to preserve civic visibility, while accounting for changed needs resulting from demographic shifts. Churches act not just to provide the rituals associated with Catholic communal life, but also as symbolic reflections of Catholicism in the city. A stated purpose gave parishes a reason to remained anchored in—but apart from—their immediate ecological contexts.

This is seen poignantly in the resurrection of St. Louis of France—"the Old Cathedral," whose grandeur and history in the city warranted a salvation rationale that superseded its diminished attendance numbers. "It is a historic church," shared an interviewee, "it was a church that was going to be retained no matter what because of its history as the oldest cathedral west of the Mississippi. And it's right by the Arch." St. Louis's impressive cathedrals and long-standing Catholic churches found new life with new purpose.

Conclusion: rootedness in the new urbanesque

Religious congregations occupy a conspicuous symbolic presence in American cities. Lending neighborhood stability (Ammerman 1997; Kinney and Winter 2006) and social capital (Farnsley et al. 2004), congregations can anchor religion in urban ecologies amidst demographic change. The relative staying power of the Catholic Church, in particular, has typically been attributed to its parishes being geographically bound (Gamm 1999; McGreevy 1996). But even a tradition reliant upon territorial designation faces the challenge of declining attendees and resource constraints necessitating church closure. Though all neighborhoods reside in a defined territorial "parish," not all parish communities thrive indefinitely.

Modern cities urge—catalyze, even—institutional innovation to maintain religion's symbolic presence, introducing a new conception of "rootedness." The new urbanesque, with its sprawl, diversity, residential mobility and cosmopolitan nature, generates religious communities less reliant upon immediate neighborhood proximity. What kept Catholicism rooted in downtown St. Louis was not the standard definition of parish as a geographic, residentially bound community. It was the intentional deployment of the personal parish as a purpose-driven, geographically boundless community of adherents. Personal parishes reconstitute ascribed "diasporic" communities across the whole metropolitan space and beyond. The Catholic Church allocated extant resources strategically to meet a rich mix of subcultures and needs. Territorial decoupling enabled Catholicism to maintain a physical presence in a changed urban setting, to affirm niche iterations of Catholic culture, and to manage divergent needs within a unified whole.

Constituting community in this way does carry the risk of isolating particular needs (for example, those related to social justice or ethnicity) to single parishes rather than to a broader, shared population of adherents. Not all needs earn a formal name or pastoral response. Personal parishes may also fragment and exacerbate divisions along ideological, racial and preferential lines. Moreover, churches that draw parishioners from far away risk neglecting the needs of residents living at the church's front steps. Parishioners' residential and financial inputs are generally directed elsewhere, furthering the economic plight of a church's neighborhood. Though beloved religious spaces may be saved, the demographic and social changes underscoring their original fate remain.

Notes

1 This research is part of a national study of personal parishes in the United States. Comprehensive findings will appear in a book by the author entitled *Parish & Place*, forthcoming with Oxford University Press.

2 Unless otherwise specified, all quotations come from 14 original, in-person, recorded interviews conducted in 2013 with St. Louis archdiocesan staff, pastors, lay pastoral

leaders and bishops who held former appointments in St. Louis. I provide minimal descriptive characteristics for specific respondents here, in line with standard sociological methods, so as to protect individuals' confidentiality. Reflecting parish and diocesan leadership, all but two interviewees were men and all but two were white.

3 The founding years of the closed parishes ranged from 1821 to 1993, averaging 1923. Technically, a parish cannot "close"; rather, another parish absorbs its territory. The church property is then sold or repurposed.

4 See Mulder (2015) for a comparison to Evangelical congregations.

References

Ammerman, Nancy T. (1997), *Congregation and Community*. New Brunswick, NJ: Rutgers University Press.

Annuario pontificio (2005), Vatican City: Librería Editrice Vaticana.

Bruce, Tricia C. (2011), *Faithful Revolution: How Voice of the Faithful Is Changing the Church*. New York: Oxford University Press.

Burke, Raymond L. (2005a), On the Pastoral Reorganization of the Northeast County Deanery. February 11, 2005. Available at http://archstl.org/files/archstl/images/stories/burke/letters2005/02-11-05-northeast_county.pdf; accessed December 1, 2014.

Burke, Raymond L. (2005b), On the Pastoral Reorganization of the South City Deanery, February 25. Available at http://archstl.org/files/archstl/images/stories/burke/columns 2005/02-25-05-column.pdf; accessed December 1, 2014.

Codex Iuris Canonici (1983), Vatican City: Libreria Editrice Vaticana.

Dolan, Jay (1985), *The American Catholic Experience: A History from Colonial Times to the Present*. Garden City, NY: Doubleday.

Farnsley II, Arthur E., N.J. Demerath III, Etan Diamond, Mary L. Mapes and Elfriede Wedam (2004), *Sacred Circles, Public Squares: The Multicentering of American Religion*. Bloomington and Indianapolis: Indiana University Press.

Gamm, Gerald (1999), *Urban Exodus: Why the Jews Left Boston and the Catholics Stayed*. Cambridge, MA: Harvard University Press.

Gordon, Colin (2008), *Mapping Decline: St. Louis and the Fate of the American City*. Philadelphia: University of Pennsylvania Press.

"History of the Archdiocese" (2001), Archdiocese of St. Louis. Available at http://archstl.org/archives/page/three-centuries-catholicism; accessed October 14, 2014.

John Paul II (1999), *Ecclesia in America*. Washington, DC: US Catholic Conference. Available at http://w2.vatican.va/content/john-paul-ii/en/apost_exhortations/documents/hf_jp-ii_exh_22011999_ecclesia-in-america.html; accessed February 12, 2015.

Kinney, Nancy T. and William E. Winter (2006), Places of Worship and Neighborhood Stability, *Journal of Urban Affairs* 28(4): 335–52.

McGreevy, John T. (1996), *Parish Boundaries: The Catholic Encounter with Race in the Twentieth-Century Urban North*. Chicago, IL: University of Chicago Press.

Mulder, Mark T. (2015), *Shades of White Flight: Evangelical Congregations and Urban Departure*. New Brunswick, NJ: Rutgers University Press.

Warner, R. Stephen (1994), The Place of the Congregation in the Contemporary American Religious Configuration, in James P. Wind and James W. Lewis (eds), *American Congregations Volume 2: New Perspectives in the Study of Congregations*. Chicago, IL: University of Chicago Press.

Webber, Melvin. M. (1963), Order in Diversity: Community without Propinquity, in Lowdon Wingo Jr. (ed.), *Cities and Space: The Future Use of Urban Land*. Baltimore, MD: John Hopkins Press.

Wedam, Elfriede (2000), Catholic Spirituality in a New Urban Church, in Lowell W. Livezey (ed.), *Public Religion and Urban Transformation: Faith in the City*. New York: New York University Press.
Weldon, Michael (2004), *A Struggle for Holy Ground: Reconciliation and the Rites of Parish Closure.* Collegeville, MN: Liturgical Press.

Part II

Producing and negotiating religious space

4 Praying in Berlin's "Asiatown"

Religious place-making in a multi-ethnic bazaar

Gertrud Hüwelmeier

Over the last two decades, large numbers of new migrants have arrived in European cities from different parts of the world. Religious imaginations and practices are part of the luggage they bring along. Therefore, religious sites such as churches, mosques, shrines, temples and other places of worship function as new markers of place-making. Their meaning and significance as linchpins in processes of diasporic religious emplacement, however, may change considerably through believers' migration and incorporation into the host society.

When it comes to religious place-making, defined as the activity of establishing a particular locality for religious practice (Hüwelmeier and Krause 2010: 8), this process of positioning goes beyond physical space alone. The new surroundings in which religious practitioners place themselves are physically bounded jurisdictions that overlap with de-localized imaginations of religious life. Thus they belong to multiple spaces at the same time. Place-making, therefore, is about the simultaneous process of being engaged globally while being situated locally. This broad definition differs from an understanding of space as a bounded entity and follows authors who look at space as socially produced (Lefebvre 1991; Massey 1993; Knott 2005; Tweed 2006). However, migrants' religious place-making in the urban landscape may lead to contestation and conflict, turning it into a delicate and complex process. This contribution seeks to explore how religious place-making emerges from the ways in which migrants transport and introduce religious ideas, practices and sacred objects from one place to another, while simultaneously changing or redefining their ideas about belief, ritual, locality and sacred space. Its particular focus is on the urban and suburban context. Based on multi-sited ethnographic fieldwork among migrants in Berlin, and in particular among Vietnamese in Berlin and their transnational ties to Hanoi, this paper focuses on the diversity of practices of religious place-making in metropolises.

Today, Buddhist performances, Catholic prayers, Pentecostal gatherings, and the veneration of gods and ancestors take place in various localities in the diaspora. Through migrant congregations, "traditional" churches, pagodas and temples as well as former warehouses, industrial buildings and marketplaces are becoming new places of worship and are thereby shaping the religious landscape in urban settings. By emphasizing the relationship between cities and transnationalism,

I argue that cities play a crucial role as key social spaces grounding transnational religious practices.[1]

Religion on the move

New migration movements, global economic orders, and communications systems have contributed to the resurgence of religion and religious practices in many parts of the world. Cities are key sites within these processes, connecting people, places and religious imaginations. Many recent religious movements, the revitalization of religion in post-socialist countries and a global explosion of Pentecostal Christianity, Hindu nationalism, transnational Islam and spirit possession cults initially emerged in urban surroundings. As cities are preferred localities for migrants, urban spaces are "at once contexts of opportunities and constraints within which transnational actors and networks operate and nodes linking wider social formations that traverse national borders" (Smith and Eade 2008: 3). This is particularly true with regard to religion, as diasporic migrants are linked to their countries of origin as well as to other places through their religious communities and practices. Religion has thrived because globalization provides fluid transnational networks that help transport religious messages from local to global audiences. But religion has contributed to those transnational networks just as much as it has been shaped by them. Rather than merely reacting to global processes, religion and religious practitioners generate global interconnectedness (Hüwelmeier and Krause 2010: 1).

In this chapter, I explore the mutually determining relationships between religious place-making and/in the city (Coleman 2009; Smit 2009; Knibbe 2009; Orsi 1999). Starting from the assumption that religion cannot be understood independently of politics, economics, consumer culture and media, this contribution contextualizes religion within transnational processes, new mobilities and migration movements. I investigate how religious practices are transformed, reinforced, or newly invented when transferred to urban spaces, how migrants rely on religion to become global actors, and how religious agents create and maintain transnational connections, simultaneously linking people and places. As a field of study, transnationalism seeks to understand how people—in single localities such as cities as well as in their movements between them—take part in multilateral national contexts through their economic, political and social activities. Such transnational "ways of being and belonging" (Levitt and Glick Schiller 2004) require abandoning a geographically fixed approach to identity and community without neglecting the continuing significance of border regimes and state policies in the control and regulation of movement (Wimmer and Glick Schiller 2002). This kind of broadened research lens is especially critical to understanding religious transnationalism, which has arisen from the conjoined processes of missionization, migration and mobility (Csordas 2007).

By focusing on the intersection between religion, space and locality (Knott 2005), I investigate how Vietnamese in Berlin appropriate urban space in order to create places of worship. In various cities around the world, migrants bring along religious practices and transform localities in the host country through place-making religious activities. Such activities can include transforming public parks into religious spaces, at least temporarily (Huang 2013), thereby shifting the urban landscape. Recent research on the conjunction of religion and urban spaces criticized the ongoing "rewriting of sociological narratives about the fate of religion in modernity" (Burchardt and Becci 2012: 1). What is still missing, however, is further research connecting an anthropology of religion with an anthropology of the city. Yet, as recent ethnographic studies have illustrated, religious place-making practices in many cities around the globe, such as the use of loudspeakers in Nigerian marketplaces by various religious agents (Larkin 2013), are of growing interest in the humanities and social sciences. This interest is likely related to the new dynamics of religious movements in metropolises and to ethnic and religious diversity in urban localities (van der Veer 2015 and 2013; Eade 2012; Hüwelmeier and Krause 2010).

Over the last ten years, I conducted ethnographic fieldwork among Vietnamese, while participating in religious rituals in various contexts, such as in Pentecostal churches and Catholic churches in Berlin and Hanoi as well as in Buddhist pagodas in both cities (Hüwelmeier 2013a) and in other places, such as Prague (Hüwelmeier 2015c) and Warsaw (2015b). I took part in spirit possession cults in Hanoi as well as in sessions with a Vietnamese fortune teller in Leipzig, a city in eastern Germany, and in Hanoi. As many Vietnamese in these locations earn their livelihood through trading, I carried out fieldwork in marketplaces in Berlin, Warsaw, Prague and Hanoi (Hüwelmeier 2013b). In addition to conducting business in these places, Vietnamese also use these spaces to perform religious practices. They simultaneously encounter Muslims, Hindus and Sikhs who are also trading in the bazaars as well as also performing prayers and other rituals in the marketplace. As a result of this situation, I refer to religious place-making practices of these other groups also present in the marketplaces as well. Further, as the lives of migrants are increasingly shaped by transnational ties and thus cannot be understood by focusing just on one city, region, or nation, I consider the transportability of religious practices (Csordas 2007) across borders, thus linking different urban spaces.

Place-making and spatial politics

Transnational mobility, media circuits and the global flow of money are grounded in, and shaped by, particular localities, such as cities or marketplaces. These spatial particularities have different positions within global hierarchies and city scales (Glick Schiller and Caglar 2011). Some positions provide privileged access to flows of power and interconnection; others require extra effort; still others create exclusion from the host society. Once established in a locality, migrants seek to practice their religion and negotiate their identities, which may involve practices of religious place-making.

Place and space are intertwined categories, which can best be explored by focusing on the hierarchization of place-making activities in urban space. The case of Vietnamese Buddhists in Berlin illustrates this kind of spatial hierarchization quite well. Before the fall of the Berlin Wall, Vietnamese boat people who had fled their home country at the end of the 1970s, after the end of the American War in Vietnam, started to establish a place of worship in a rented apartment before purchasing a house with garden in a purely residential area in Berlin Spandau, a north-western city district, in the 1980s. When I started my ethnographic fieldwork on transnational religious networks of Vietnamese in 2006, I visited this locality several times. Two years later, the place no longer existed. The Vietnamese Buddhist community sold the house and garden and purchased a former office building near a commercial estate also in Berlin Spandau in an area that was used as a production site for the arms industry in the nineteenth century and during World War I.

In 2011, the community started building a pagoda on the grounds of the office building. The new pagoda was inaugurated in 2012 and since then this site has been used for religious rituals by Vietnamese boat people, most of them anticommunists, and by Vietnamese former contract workers from socialist East Germany, most of them still loyal to the communist regime in Vietnam.

The practice of renting non-sacred spaces and repurposing them for religious rituals occurs among migrant groups in other German cities as well. In Hanover, a city in western Germany, Vietnamese refugees rented a former laundry in the 1980s, and later moved on to a small apartment to perform religious practices. Years later, they purchased a plot of land outside Hanover to build one of the largest Vietnamese Buddhist pagodas in Europe, near the highway and far away from the city center (Baumann 2000). This corresponds to the trajectory of many Muslims in Berlin who founded storefront mosques in the 1970s and 1980s by renting small places in order to establish prayer rooms. Later they moved to bigger places as more space was needed to accommodate the growing congregations. Between 1995 and 2005, a kind of consolidation of mosques and Islamic prayer rooms took place in Berlin. This can be observed from the fact that approximately one-quarter, namely 21 of the registered Muslim organizations in 2005, have become owners of former industrial lofts, building plots, front buildings, side wings and rear wings, or whole floors in residential houses in order for their members to gather religiously (Färber and Spielhaus 2010: 100). Thus, there seems to be no longer a noticeable mobility with regard to the Muslim sacred landscape in Berlin. There is rather a kind of religious settlement in the urban environment. Today, five impressive mosques encompass the visibility of Islamic places in Berlin, and, moreover, about eighty neighborhood prayer rooms and a number of temporarily used places are shared and shaped by believers, such as spaces where adherents of Sufi Islam or Alevis perform their rituals on special occasions. With regard to Buddhists and Muslims alike, processes of becoming visible can be observed in the typical spatial career of a migrant's religious network: At first, a few people start out meeting in a private space; eventually they pay for a location on an hourly basis; then they rent a place entirely for themselves, and finally they buy or build their own place of worship. However, these processes are not always without tensions and conflicts within the respective religious community and in the host society.

Controversies and disturbances

Religious place-making brings major challenges with it. Differing concepts of the religious and the secular and the power relations through which public space becomes articulated (Asad 2003: 184) determine the availability of space for religious veneration. As a result, spatial politics always go hand-in-hand with place-making. Sacred sites function as markers in the urban landscape, reflecting the social position of the respective religious group. It should therefore come as no surprise that place-making can erupt in conflict—the controversies surrounding mosques in Europe are only recent examples. The case of a mosque in Berlin-Heinersdorf, a neighborhood in eastern Berlin's Pankow district, is quite challenging for politicians, neighbors and religious practitioners as well. A citizens' action committee tried to impede the foundation of the Khadija Mosque, which is part of the Ahmadiyya Muslim Jamaat, an Islamic reform movement. However, they were not successful and, since the fall of the Berlin Wall, the first mosque with a minaret in eastern Germany was inaugurated in 2008. This building is located outside the city's center, near a slip road and therefore, like many other religious places of migrants, on the margins of the city.

Since space in inner-city areas is more expensive and new religious groups often have difficulties obtaining a lease, some move to the urban outskirts, where they acquire former warehouses, or rent garages to use for religious gatherings. Many Pentecostal Christians, for example, appreciate the practical advantages offered by former industrial buildings for their mega-churches. Here they can easily seat large crowds, employ elaborate media equipment, and display the power of the Word on a huge stage (Coleman 2000: 155). For many small churches founded by migrants in Europe, the most important practicality of industrial areas is the fact that nobody cares about the sound-system-enhanced praying and preaching that accompanies their ritual practice, as Kristine Krause has analyzed for the city of London (Krause 2008). With regard to prayer and religious sound-making practices of Muslims in urban spaces as mentioned above, publicly broadcast cassette sermons in Cairo (Hirschkind 2006), for example, and other practices involved in the making of Muslim sacred places in cities (Desplat and Schulz 2012) illustrate the new audibility of religious messages in the public sphere.

Noise is a crucial issue when religious practitioners gather in certain neighborhoods. While I was participating in the prayer meetings of the Vietnamese Pentecostal Holy Spirit Church[2] in eastern Berlin, church members first closed the windows of the rented space, in order not to disturb the neighbors on a Sunday afternoon. The sound system is an essential piece of equipment for casting out demons and thus purifying the space (Hüwelmeier 2014). Loud singing is part of the religious service and hence considered part of prayer. In the urban environment, this kind of bodily expression might cause problems. This issue was exemplified by the situation of a house church of Pentecostal Vietnamese in Berlin, whose religious gatherings I attended for several months in 2010. The group consisted of about thirty female followers who together split off from another Pentecostal church and then met in the home of its leader, Mrs. Tung.

She told me she was very anxious about neighbors complaining about the noise of loud singing and praying coming from her rented apartment. She had been living in the multi-story building for several years with her husband and three children, and she was worried that official complaints from the people living nearby could lead to the termination of her rental agreement. One day she told me:

> You know, it is not at all easy for a Vietnamese family to find such a nice apartment with a balcony in the centre of the city, most Vietnamese live in the suburbs. But as God told us to leave the other group, we need a place to praise the Lord. When we first met in my home several years ago, we even used my bathtub to perform baptism.

As Vietnamese, like many migrant groups, face great difficulties in finding a venue for religious gatherings, they try to avoid tensions with the neighbors and the venue's owner. This is one reason why migrant congregations change their religious localities several times within short periods of time, often moving around the city.

Transforming secular and religious localities

Religious practitioners not only construct new sacred places, but can also transform existing secular localities into religious ones (Garbin 2013). This results in spaces of leisure and consumption such as cinemas, bowling alleys and discotheques being turned into shrines, temples and churches (Meyer 2006: 295; Ukah 2003). In the case of a Vietnamese Pentecostal church in northern Germany that I visited several times during my fieldwork, followers first rented space on an hourly basis in a free evangelical church in the city of Bremen. However, after having established a stable group of Christian believers, the community decided to purchase a building outside the city, a former country inn with a bowling alley. The bowling alley is no longer used for recreation; it has been transformed into mass sleeping accommodations for the church's annual convention. For this event, the space hosts about 300–400 people from all over Germany as well as guests from Latin America and Norway, as the church has close connections to Pentecostal networks abroad. This example illustrates that there is not only movement from rural areas into cities in order to find places of worship, but also the other way round (Hüwelmeier 2010: 221).

During my ethnographic fieldwork, I realized that several Pentecostal Vietnamese churches were moving repeatedly to different sites in eastern Berlin, as, among other reasons, the rental fee for the rooms they wanted to use was too high. In one case, a church gathered in a community center, right in the middle of an estate of multi-story pre-fabricated apartment blocks (*Plattenbausiedlung)*, an architectural type that had been part of socialist planning (Hirsh 2013). Most of the church members lived in this neighborhood. Another group of charismatic Pentecostal Vietnamese gathered on the grounds of a wholesale bazaar in Berlin

Marzahn, a former industrial area in the eastern part of the city, with the aim of attracting traders who had not yet found their way to Jesus. As these examples illustrate, it is not always the inner city that is attractive for migrants' religious gatherings. In many cases, backyards, storefront mosques, living-room churches, or warehouse pagodas are the spaces on the outskirts of big cities where religious practitioners become invisible or are made invisible to the surrounding society, as the inner city does not provide a dedicated place of worship that these congregations can access.

Praying in "Asiatown"

In urban spaces, there is a considerable number of groups whose spatial visibility, at least temporally, involves a movement from private to public space, and from invisibility to visibility, for example, via parades (Goreau 2014), or, as mentioned above, the use of loudspeakers by religious agents in Nigerian marketplaces (Larkin 2013). Simultaneously, there are many religious practitioners who do not strive for public recognition, or, depending on the religious majority of a country, do not feel safe in the city, as Heck has discussed for African Christians in Istanbul (Heck 2013). In Berlin, a newly built mosque or a Buddhist pagoda is quite visible in the cityscape due to its impressive architecture. However, there are many other places in the city, invisible to the outsider at first glance, where migrants perform prayers, venerate gods, or worship ancestors.

Among these localities are Asian marketplaces, run by migrants that are temporarily transformed into sacred spaces. In order to understand why some adherents perform religious rituals in particular localities such as in a marketplace, one should focus on the kind of religious practices that occur, on the various groups of migrants who use such spaces, and on the reasons why people pray in their places of work. In my recent ethnographic fieldwork on "global bazaars," in particular on Asian wholesale markets in Berlin, I witnessed a close connection between the economy and religion. As Max Weber has analyzed more than a hundred years ago in his well-known book *The Protestant Ethic and the Spirit of Capitalism* (Weber 2002 [1905]), the economy and religion are not separate spheres, but intrinsically interwoven on various levels. Economic success is a crucial issue for today's migrants, and requests for such success are often part of the prayers in pagodas, churches, mosques and places of business as well. Offerings and donations play an important role in these religious rituals and are used to underline the request for support from spiritual entities. As the desire to do good business is a driving force in performing religious practices for many people, particularly migrants, it is not surprising that marketplaces, which form an important part of the urban landscape, are localities where prayer is performed in order to obtain the protection and the blessings of gods and spirits.

Unlike the marketplaces investigated in Britain (Watson 2009), new Asian bazaars in Berlin are not public places, but private sites, owned by Vietnamese who had once been employed as contract workers in East Berlin before the fall of the Wall. It was only many years after the fall of the Wall in 1989 that a migrant investor purchased a former industrial site and built a new place called the Dong Xuan Center, a wholesale market in Berlin Lichtenberg run by Vietnamese and traders from various countries such as India and Pakistan.

The market was newly constructed in 2005 in a bleak industrial area whose architecture is characterized by many pre-fabricated houses from the socialist period. Today, this area is home to about 4,000 Vietnamese. With its cheap products mainly from China, its Vietnamese restaurants and Asian supermarkets, this market is a perfect place for meeting other people or just looking around. Various other migrant groups live in this area as well, such as people from Poland, Russia, Serbia and Montenegro. Some of them visit the Dong Xuan Center on a regular basis. Furthermore, Germans living in the neighborhood, a number of them poor and unemployed, like to go shopping in the multi-ethnic market and visit the Vietnamese tattoo shop or the inexpensive Vietnamese hairdresser in one of the large halls. Recently, the center became an insider tip for many Berlin tourists.

This locality was previously considered the property of the socialist government of the German Democratic Republic. Today, this site is privately owned. It has also developed into a multi-ethnic as well as a multi-religious place. This is particularly clear when one considers that the new business people and market managers invite not just retailers from all over Germany, Poland, the Czech Republic, the Netherlands and Denmark, and from regions such as Asia, Africa and

Figure 4.1 Entrance to the Dong Xuan Center in Berlin. Altars are to be found in many of the stalls in the center. Photo by Gertrud Hüwelmeier.

Latin America to visit the Asian bazaar in eastern Berlin, but also welcome tourists and inhabitants from the surrounding neighborhood.

Various religious practices are performed by people working in the market, depending on the political, ethnic and religious background of traders. For example, I noticed a Turkish trader ask another man, whom he knew personally, to keep an eye on his business while he retreated into the rear corner of his retail area in order to perform his prayers on a prayer rug. Meanwhile, the business went on, clients entered the stall, looking around, asking for prices, bargaining, buying and leaving again. Nobody took notice of the Muslim kneeling down to fulfill his religious duties. Only recently has a mosque been established next to the market. Muslim traders of various backgrounds told me that they no longer have to travel back and forth to city mosques in western Berlin to perform prayers during their working hours.

Some traders in the bazaar are Sikhs from India, who, in their stalls, near the cashbox, display images of their main Gurdwara (prayer site) in Amritsar, the Golden Temple, in Punjab, Northern India, near the border to Pakistan. Sikhs in the Asian marketplace pray twice a day in their workplace, but only if time allows during busy trading. A Sikh trader in his forties said, "I feel my business being protected by the Golden Temple." According to him and some other Sikh traders, the spirituality of the sacred place Amritsar has a positive influence on their trading activities and everyday lives in the Dong Xuan Center. The trader explained to me, "Look, the Golden Temple is always part of my life. I show you some photos

Figure 4.2 "Asiatown": worship of the spirits of the place, Muslim prayers and ceremonies of Buddhists take place in the stalls of traders. Photo by Gertrud Hüwelmeier.

of the temple." Then he took out his smartphone and showed video-clips about the massacre in the holy city of the Sikhs in 1919, ordered by the British. He also demonstrated videos about the removal of Sikh militants from the Golden Temple in Amritsar in 1984, an Indian military operation ordered by Indira Gandhi, which finally led to her assassination. The use of new media technologies illustrates how people from various parts of the world transport religious messages and how religion and technology are intrinsically interwoven, connecting people and places across borders (Hüwelmeier 2015a; Behrend et al. 2015; Meyer 2015).

Similar to Muslims and Sikhs, shop owners of Vietnamese background perform religious rituals in the marketplace. Recently, a pagoda was inaugurated on the grounds of the Asia-Pacific Center, another wholesale market in Berlin Marzahn, near the Dong Xuan Center. I participated in the inauguration ceremony in 2007, which took place in a former porter's office that had been used by the East German secret service before 1989. The space was purchased by Vietnamese business people in 2003 and the porter's office was transformed into the office of the new market owners, who were former contract workers from Vietnam. The manager, a woman in her fifties, explained to me:

> When I was young, there was war in Vietnam. I was lucky and came to eastern Germany as a contract worker in the late 1980s. Then the Wall came down and again I was lucky and started a wholesale business. I became rich and so I feel it is my duty to share some of my wealth with other people. Therefore I leave part of my office to the lay Vietnamese Buddhist Community, some of them are traders and clients in the market. It is very good for the traders' business, when they pray before they start working in the market.

The "market pagoda" in the Asia Pacific Center was established by a group of Buddhist lay practitioners, who renovated the porter's office while simultaneously creating a garden with flowers and Buddha statues near the entrance of the former secret service location. Several Buddha statues were purchased by the market manager in Ho Chi Minh City and shipped to Berlin, where they arrived only a few days before the inauguration ceremony was scheduled to take place. As this case illustrates, religious objects are shipped from the global South to the global North, from the East to the West, and are hence part of what has been termed "traveling religions" relying on portable practices (Csordas 2007: 261). At the start of the inauguration ceremony, the Buddha statues were still packed in polythene sheeting and wooden cases, waiting to be unpacked and placed in the inner space of the pagoda. It is not only religious objects that travel for religious purposes, but also religious experts. The inauguration of a pagoda requires that a ritual be carried out by a monk. Thanks to contact between Vietnamese Buddhist practitioners in Berlin and France, the Berlin community was able to invite a monk from a Buddhist monastery near Paris to perform the ritual.

A sacred space within an Asian wholesale market in Berlin does not seem to be unusual in the eyes of Vietnamese clients and visitors. In marketplaces in Hanoi, traders ask for spiritual guidance in the nearby pagodas of the ancient town, praying for a successful business, and a happy family. Likewise in Berlin, before they enter the bazaar, clients pray for health and wealth in the market pagoda of the Asia Pacific Center. Further, customers pray for protection with regard to entrepreneurial activities and for success in education for their children. Usually, they bring offerings such as food and light incense. In the case of a need, such as marital problems, for example, Vietnamese consult the monk who lives on the grounds of the market pagoda. Traders as well as clients appreciate the inauguration of the pagoda in the Asia Pacific Center, which was the first one to be established in eastern Berlin after the fall of the Wall. Another Vietnamese Buddhist pagoda, situated in western Berlin, which I mentioned above, is quite far

Figure 4.3 Ban tho (altar) in a Vietnamese shop, Don Xuan Center, Berlin. Photo by Gertrud Hüwelmeier.

from the trade centers in the eastern part of the city. Thus, as most of the traders and business people do not have time to go there on a regular basis, they prefer to pray in the bazaar pagoda (Hüwelmeier 2013a).

In line with the practices of Buddhists in contemporary socialist Vietnam (Taylor 2007), Vietnamese shopowners in Berlin not only visit the pagoda, but also perform various religious rituals in marketplaces. For example, Vietnamese business owners establish small altars (*ban tho*) near the cashbox, and make offerings of fruit and alcohol to the God of Wealth (*Ong Than Tai*) and the God of the Earth (*Ong Dia*) in their salesrooms (Hüwelmeier 2008: 140), just as they do in the markets of Hanoi and Ho Chi Minh City (Leshkowich 2014:161) In the Dong Xuan Center in Berlin, traders purchase these small altars, which are shipped from Vietnam and are on display in the food section of the Asian supermarket. In their respective stalls, the owner and his or her family decorate the altar with small figures, representing gods, bring offerings such as vodka, rice and cigarettes, and pray every morning before starting business.

Praying for successful trading, good health, and a happy family is quite important in many Vietnamese families who are engaged in commercial transactions. Therefore, "Asiatown," as the Dong Xuan Center is conceptualized in a future project of the market manager, can thus be identified as an urban locality where multi-ethnic groups share a common space, not just for trading, but also for religious purposes.

Conclusion

As investigated in this chapter, the re-enchantment of the urban landscape and the performance of religious practices by migrants are manifested not only through religious place-making activities on the ground, but also through practitioners' global outreach. Cities are arenas for exploring the nexus of cross-border religious networks and their impact on place and vice versa (Smith 2001: 183). They are thus simultaneously interconnected sites for the emplacement of mobile people, rituals and objects. Sacred architecture, prayer, images, statues and movies circulate between various localities, cities, regions and nations.

Anthropological fieldwork has illuminated the close connection between religious place-making practices and the city. In particular with regard to the history of Berlin, its division into West and East, and the arrival of different groups of Vietnamese before and after the fall of the Wall, illustrate the tight conjunction between sacred places, and the transformation of secular localities and urban space. By focusing on the geographic mobility of migrants across the urban landscape in pursuit of spaces to conduct religious activities, this chapter has indicated how and why migrants frequently change locales for worshipping. Therefore, not only the fluidity and pace of the city, but in particular the economic constraints and opportunities of urban life shape the typical spatial career of migrant religious networks, looking for better and affordable localities in order to perform prayer and other religious duties. Since space in the inner-city areas is too expensive for migrant religious networks, some of them move to urban outskirts, where they acquire former warehouses, office buildings, or wholesale markets in suburbia.

The hierarchization of place-making activities ranges from groups meeting in a private living room to finally buying or building their own place of worship. Migrant groups' frequent change of worship site and the accompanying processes of sacralizing and de-sacralizing of locations highlight the temporality of the religious nature of places, which are oftentimes no longer considered sacred after a religious group has left to appropriate another locality.

The ethnographic case study of Berlin's "Asiatown" has illustrated that diverse ethnic groups perform different religious practices in a seemingly secular, however urbanesque, environment, namely on the grounds of a former industrial area, today a global trading place: a wholesale bazaar on the city's eastern periphery. By transporting religious objects to Berlin, by inviting religious experts from abroad and by praying in the marketplace, migrants at once recall the places of worship in their countries of origin and emphasize the transnational ties between different urban spaces. The trajectory of Vietnamese Pentecostal and Buddhist worshippers in Germany is in many ways typical for migrant religious groups in an urban landscape, while simultaneously presenting unique aspects arising from the varied political and socio-economic differences between Vietnamese living in a country still marked by its decades of division.

Notes

1 This contribution is based on results of my ongoing research project "The Global Bazaar," funded by the German Research Foundation (HU 1019/3-1), and on my former research project on "Transnational Religious Networks" (HU 1019/2-1 and HU 1019/2-2), funded by the German Research Foundation as well. First thoughts on religious place-making were presented in the introduction of a workshop on religious emplacement I organized with Rijk van Dijk at the EASA (European Association of Social Anthropologists) conference in Bristol, UK, in 2006. Further considerations on religious place-making came up during a conference in Kraków on transnational ties and city space, organized by John Eade and Michel Peter Smith in 2007 (Smith and Eade 2008). That same year I co-organized a conference on "Travelling Spirits" at the Humboldt University in Berlin, where the relationship on place, space and religious practices was discussed in various presentations (Hüwelmeier and Krause 2010: 7–8).

2 This name is a pseudonym.

References

Asad, Talal (2003), *Formations of the Secular: Christianity, Islam, Modernity*. Stanford, CA: Stanford University Press.

Baumann, Martin (2000), *Migration-Religion-Integration: buddhistische Vietnamesen und hinduistische Tamilen in Deutschland*. Marburg: Diagonal-Verlag.

Behrend, Heike, Anja Dreschke and Martin Zillinger (eds) (2015), *Trance-Mediums and New Media*. New York: Fordham University Press.

Burchardt, Marian and Irene Becci (2012), Religion Takes Place. Producing Urban Locality in Irene Becci, Marian Burchardt and Jose Casanova (eds), *Topographies of Faith: Religion in Urban Spaces*. Leiden: Brill, 1–21.

Coleman, Simon (2000), *The Globalisation of Charismatic Christianity: Spreading the Gospel of Prosperity*. Cambridge: Cambridge University Press.

Coleman, Simon (2009), The Protestant Ethic and the Spirit of Urbanism, in Rik Pinxten and Lisa Dikomitis (eds), *When God Comes to Town: Religious Traditions in Urban Contexts*. London: Berghahn, 33–44.

Csordas, Thomas J. (2007), Introduction. Modalities of Transnational Transcendence, *Anthropological Theory*, 7(3): 259–72.

Desplat; Patrick A. and Dorothea E. Schulz (eds) (2012), *Prayer in the City: The Making of Muslim Sacred Places and Urban Life*. Bielefeld: Transcript Verlag.

Eade, John (2012), Religion, Home-making and Migration across a Globalising City: Responding to Mobility in London, *Culture and Religion: An Interdisciplinary Journal*, 13(4): 469–83. London: UCL/University of Roehampton.

Färber, Alexa, and Riem Spielhaus (2010), Zur Topografie Berliner Moscheevereine. *Berliner Blätter*, 53: 96–111.

Garbin, David (2013), The Visibility and Invisiblity of Migrant Faith in the City. Diaspora Religion and Politics of Emplacement of Afro-Christian Churches, *Journal of Ethnic and Migration Studies*, 39(5): 677–96.

Glick Schiller, Nina and Ayse Caglar (eds) (2011), *Locating Migration: Rescaling Cities and Migrants*. Ithaca, NY: Cornell University Press.

Goreau, Anthony (2014), Genesha Carturthi and the Sri Lankan Tamil Diaspora in Paris. Inventing Strategies of Visibility and Legitimacy in a Plural Monoculturalist Society, in Ester Gallo (ed.), *Migration and Religion in Europe: Comparative Perspectives on South Asian Experiences*. Farnham: Ashgate, 211–31.

Heck, Gerda (2013), Worshiping at the Golden Age Hotel. Transnational Networks, Economy, Religion, and Migration of the Congolese in Istanbul, in Jochen Becker, Katrin Klingan, Stephan Lanz and Kathrin Wildner (eds), *Global Prayers: Contemporary Manifestations of the Religious in the City*. Berlin: Metrozones, 274–89.

Hirsh, Max (2013), Intelligentsia Design and the Postmodern Plattenbau, in Kenny Cupers (ed.), *Use Matters: An Alternative History of Architecture*. London: Routledge, 169–82.

Hirschkind, Charles (2006), Cassette Ethics: Public Piety and Popular Media in Egypt, in Birgit Meyer and Annelies Moors (eds), *Religion, Media, and the Public Sphere*. Bloomington: Indiana University Press, 29–51.

Huang, Weishan (2013), The Geopolitics of Religious Spatiality and Falun Gong´s Campaign in New York, in Irene Becci, Marian Burchardt, and Jose Casanova (eds), *Topographies of Faith: Religion in Urban Spaces*. Leiden: Brill, 129–49.

Hüwelmeier, Gertrud (2008), Spirits in the Market Place of Vietnamese Migrants in Berlin, in Michael Smith and John Eade (eds), *Transnational Ties: Cities, Identities, and Migrations*. New Brunswick, NJ: Transaction Publishers, 131–44.

Hüwelmeier, Gertrud (2010), Dämon oder Holy Spirit? Geistbesessenheit in einer vietnamesisch-pentekostalen Gemeinde in Berlin, in Dorothea Schulz and Jochen Seebode (eds), *Spiegel und Prisma: Ethnologie zwischen postkolonialer Kritik und Deutung der eigenen Gesellschaft*. Hamburg: Argument Verlag, 203–16.

Hüwelmeier, Gertrud (2013a), Bazaar Pagodas – Transnational Religion, Postsocialist Marketplaces and Vietnamese Migrant Women in Berlin, *Religion and Gender*, 3(1): 75–88.

Hüwelmeier, Gertrud (2013b) Post-socialist Bazaars. Diversity, Solidarity and Conflict in the Marketplace, *Laboratorium. Russian Review of Social Research*, 5(1): 42–66.

Hüwelmeier, Gertrud (2014), Performing Intimacy with God: Spiritual Experiences in Vietnamese Diasporic Pentecostal Networks, *German History*, 32(3): 414–30.

Hüwelmeier, Gertrud (2015a), New Media and Traveling Spirits. Pentecostals in the Vietnamese Diaspora and the Disaster of the *Titanic*, in Heike Behrend, Anja Dreschke and Martin Zillinger (eds), *Trance-Mediums and New Media*. New York: Fordham University Press, 100–115.

Hüwelmeier, Gertrud (2015b), From "Jarmark Europa" to "Commodity City". Socialist Pathways of Migration and the Creation of Transnational Economic, Social and Religious Ties, *Central and Eastern Migration Review*, 4(1): 27–39.

Hüwelmeier, Gertrud (2015c), Mobile Entrepreneurs. Transnational Vietnamese in the Czech Republic, in Hana Cervinkova, Michał Buchowski and Zdeněk Uherek (eds). *Rethinking Ethnography in Central Europe*. London: Palgrave Macmillan, 59–73.

Hüwelmeier, Gertrud and Kristine Krause (eds) (2010), *Traveling Spirits: Migrants, Markets, and Mobilities*. Oxford and New York: Routledge.

Knibbe, Kim (2009), "We Did Not Come Here as Tenants, But as Landlords": Nigerian Pentecostals and the Power of Maps, *African Diaspora*, 2(2): 133–58.

Knott, Kim (2005), *The Location of Religion: A Spatial Analysis*. London: Equinox.

Krause, Kristine (2008), Spiritual Spaces in Post-industrial Places: Transnational Churches in North East London, in Michael Peter Smith and John Eade (eds), *Transnational Ties: Cities, Identities, and Migrations*, CUCR book series 9, 109–130.

Larkin, Brian (2013), Techniques of Inattention. The Mediality of Loudspeakers in Nigeria, in Jochen Becker, Katrin Klingan, Stephan Lanz and Kathrin Wildner (eds), *Global Prayers: Contemporary Manifestations of the Religious in the City*. Berlin: Metrozones, 352–67.

Lefebvre, Henri (1991), *The Production of Space*. Oxford: Blackwell.

Leshkowich, Ann Marie (2014), *Essential Trade: Vietnamese Women in a Changing Marketplace*. Honolulu: University of Hawai'i Press

Levitt, Peggy and Nina Glick Schiller (2004), Transnational Perspectives on Migration: Conceptualizing Simultaneity, *International Migration Review*, 38(145): 595–629.

Massey, Doreen (1993), Questions of Locality, *Geography*, 78: 142–9.

Meyer, Birgit (2006), Impossible Representations: Pentecostalism, Vision, and Video Technology in Ghana, in Birgit Meyer and Annelies Moors (eds), *Religion, Media, and the Public Sphere*. Bloomington, Indiana: Indiana University Press, 290–312.

Meyer, Birgit (2015), *Sensational Movies: Video, Vision, and Christianity in Ghana*. Oakland: University of California Press.

Orsi, Robert Anthony (1999), *Gods of the City: Religion and the American Urban Landscape*. Bloomington: Indiana University Press.

Smit, Regien (2009), The Church Building: a Sanctuary or a Consecrated Place? Conflicting views between Angolan Pentecostals and European Presbyterians, *African Diaspora*, 2(2): 182–202.

Smith, Michael Peter (2001), *Transnational Urbanism: Locating Globalization*. Malden MA: Blackwell Publishers, 1–20.

Smith, Michael Peter and John Eade (eds) (2008), *Transnational Ties: Cities, Identities, and Migrations*. New Brunswick, NJ: Transaction Publishers.

Taylor, Philip (ed.) (2007), *Modernity and Re-enchantment in Post-revolutionary Vietnam*, Singapore: ISEAS Publishing Institute of Southeast Asian Studies, 342–70.

Tweed, Thomas A. (2006), *Crossing and Dwelling: A Theory of Religion*. Cambridge, MA: Harvard University Press.

Ukah, Asonzeh (2003), Advertizing God: Nigerian Christian Video-Films and the Power of Consumer Culture, *Journal of Religion of Africa*, 33(2): 203–31.

Van der Veer, Peter (ed.), (2015). *Handbook of Religion and the Asian City: Aspiration and Urbanization in the Twenty-first Century*. Berkeley: University of California Press.

Van der Veer, Peter (2013), Urban Aspirations in Mumbai and Singapore, in Irene Becci, Marian Burchardt and Jose Casanova (eds), *Topographies of Faith: Religion in Urban Spaces*. Leiden: Brill, 61–73.

Watson, Sophie (2009), The Magic of the Market Place: Sociality in a Neglected Space, *Urban Studies* 46(8), 1577–1591.

Weber, Max (2002) [1905], *The Protestant Ethic and the Spirit of Capitalism*, translated by Peter Baehr and Gordon C. Wells. London: Penguin Books.

Wimmer, Andreas and Nina Glick Schiller (2002), Methodological Nationalism and Beyond: Nation-state Building, Migration and the Social Sciences, *Global Networks*, 2(4): 301–34.

5 Framing the Pope within the urban space

John Paul II and the cityscape of Kraków

Anna Niedźwiedź

Introduction

In this chapter, I will discuss the relationship between the urban space of Kraków, located in southern Poland, and the development in that city of the religious cult of the late Pope John Paul II, a canonized saint of the Catholic Church since April 27, 2014. Even though Karol Wojtyła was born in the small town of Wadowice and moved to Kraków only in 1938, at the age of 18, in Poland today Kraków is widely known as "the Pope's City." My special focus will be on how this city is *lived* as a space related to John Paul II, and how it reveals the various identities of people who use and create this urban space. Acknowledging that "city folk do not live *in* their environments; they live *through* them" (Orsi 1999: 44) and that the religiously and spiritually important places might be "used as markers of identity" (Eade and Katić 2014: 8), I propose to enumerate three discourses related to John Paul II, which are framed within the cityscape of Kraków. Provisionally, I distinguish these discourses as "national," "local" and "global"; in the subsequent parts of this chapter, I will discuss them in detail. Such discourses might be treated as parts of the "urban imaginary" and the local "urban mindscape," understood as a "mental construct of the city" and a "structure of thinking about a city" (Bianchini 2006: 13–16). Religion and the Pope are included in Kraków's (sub)urban imaginary. Popular discourses about the Pope are framed within the cityscape, while simultaneously forming and influencing how the city is lived by people, constituting new forms of devotionalization and spiritualization.

The ethnographic material on which this study is based was collected among Kraków's inhabitants and visitors to the city. Specifically, ethnographic participant observation and interviews were conducted during the period of mourning after the death of the Polish Pope on April 2, 2005, and subsequently in November 2005 and in 2008, as well as during the annually organized pilgrimages "In the Footsteps of Karol Wojtyła the Worker," in 2007, 2009, 2012, and 2013. However, the main body of the material consists of 200 interviews with Polish people who identify themselves as Roman Catholics, albeit to varying degrees, conducted by a team of students I supervised in 2013–14. These interviewees were visitors to a huge, modern and recently developed devotional complex located in the southern suburban part of Kraków called Łagiewniki.[1] The complex in Kraków-Łagiewniki consists of the Divine Mercy shrine with its immense basilica (consecrated by John Paul II in 2002) and the shrine of John Paul II, built after the Pope's beatification on May 1, 2011.

Focusing on one specific urban case study and relating it to a specific religious tradition, this analysis reveals, however, more generally observable dynamics between a city and its people. I suggest that, on the one hand, urban space appears to be an important subject of the activities of those who shape and transform it. On the other hand, urban space can be treated as an agent in itself, an entity, which generates, influences, or even imposes certain discourses on people who use this space while living within it, visiting it, or simply passing through it. As pointed out by Thomas A. Tweed, "space" can be treated as "*differentiated*, *kinetic*, *interrelated*, and *generative*" (Tweed 2011: 117, emphasis in original). Recognizing the place-making power of contemporary urban Christianities, emphasized by James S. Bielo, who focuses on "Christian modes of urban experience" and asks "how do Christian communities cultivate senses of place that directly reflect and recreate their social and religious lifeworlds" (Bielo 2013: 301, 304), I realize that, space, which is *lived*, establishes itself in people's experiences and is inscribed in people's memories and affective discourses (Duff 2010). Thus, the city and its cityscape are produced, but also productive (Eade and Mele 2002: 8). Here, I am interested in how this two-way productivity operates in a religious context. Further, I explore how it involves the popular figure of a contemporary saint, perceived in Poland both as a national hero, and as a real person, someone remembered by people and treated familiarly. As revealed during ethnographic fieldwork, the (sub)urban scape of Kraków reflects and generates various discourses on the Pope that are represented, established, reaffirmed and lived through spatial and architectural material forms, narratives and memories as well as through peoples' activities and rituals, performed in particular places located in the city.

The "nationalized" city and the Polish Pope

The Pope's "grave" in Kraków

In a small, dark chapel, dimmed light is focused on a massive white marble tombstone exposed on the floor, in the central area of a rectangular interior, in front of a tabernacle and a simple table altar. Two rows of concrete columns joined by arches divide the tiny, windowless chapel into three naves. Both sidewalls are covered with glass cabinets containing a peculiar exhibition: a white papal robe, old shoes, and a wooden crucifix. A few people kneel in front of the tombstone. Some come closer, genuflect, touch the stone and kiss a golden reliquary attached to its head. The candlelight reaching the stone from nearby illuminates simple block letters engraved in the marble stone: JOANNES PAULUS PP II, 16 X 1978 – 2 IV 2005. A small printed note in front of the stone informs that "here is the blood relic and the tombstone from the Vatican grave of St. John Paul II."

This chapel was constructed in 2011 in the basement of an immense, newly built shrine dedicated to John Paul II as a part of modern religious complex

Figure 5.1 People praying in front of the symbolic "Pope's grave" in the St. John Paul II shrine, Kraków-Łagiewniki, September 2013. Photo by Anna Niedźwiedź.

in Kraków-Łagiewniki—a southern suburban area outside the proper city of Kraków. (The lower part of the shrine was officially erected in June 2011, the upper church in June 2013) The small chapel with the Pope's tombstone is called the "Priesthood Chapel." It is designed as a replica of the Romanesque Crypt of St. Leonard, located in a vault of the Wawel Royal Cathedral in the center of town. The original crypt is directly connected with Kraków's part in the biography of John Paul II. This is where, in 1946, at the dawn of Communist, post-World War II Poland, Karol Wojtyła said his first mass as a newly ordained Catholic priest. Later, when he became the archbishop of the diocese of Kraków, the Wawel Cathedral served as his Episcopal church until 1978, when he was elected Pope and moved to the Vatican.

Thus, the design of the new location for the Pope's original Vatican tombstone is by no means accidental. This tombstone was transported to Kraków following the beatification of John Paul II in 2011. Before the beatification ceremony, the casket containing the body of John Paul II was exhumed and moved from its original grave in the Vatican crypts to be reinterred in the Chapel of St. Sebastian in St. Peter's Basilica. The original tombstone from the Vatican crypt was subsequently donated to the city of Kraków. It was transported to Poland and installed in the newly built shrine in Kraków-Łagiewniki within a space replicating the Crypt of St. Leonard from the Wawel Cathedral in Kraków's Old Town. For Polish visitors to this new shrine of John Paul II in Kraków-Łagiewniki, the local "grave" of the Pope constitutes a spatial and material form, which has religious-national connotations. Grzegorz Brzozowski suggests that the chains of memory evoked by the figure of the Pope in Poland are intertwined: they are simultaneously religious and national (Brzozowski 2013: 247). Here, agreeing with Brzozowski, I will explain the religious and national connotations associated with the Pope and the cityscape of Kraków.

Religious-national discourses in the ethnographic material

John Paul II—referred to by our interlocutors usually as simply "the Pope" or "our Pope"—is frequently described in religious-national terms. People talk about him as a holy figure, a saint and religious leader of extremely high spiritual qualities. A few chosen statements collected during fieldwork vividly represent these convictions: "He is a saint, but he lived in our times, in front of our eyes."[2] "We surely believed in his special divine dimension when he was here, on earth, even though the miracles, which happened during his life had not been spoken of aloud. John Paul II lived next to me, and I could see that his life was holy."[3] "He was really a unique person. There will be no one else like him."[4] "You could say that he is the second figure after Jesus, the holiest man! . . . He was teaching as Jesus once taught. He was revealing the same words Jesus revealed. For me he is a holy man."[5]

Simultaneously, many people emphasize the Pope's Polishness and praise his role as a national hero and a leader of the Poles, one who instigated political changes in East-Central Europe: "He expelled Communism . . . he knocked it down."[6] "For Poles he was a symbol of freedom. That is why we treat him as our Pope."[7] "He was a very important historic figure related to the fall of Communism in Poland."[8] "He influenced us Poles, so we raised our heads, we felt stronger, freer. He woke up our emotions, and they contributed to the creation of Solidarity [the Solidarity movement, in 1980] and to liberation from Communist occupation."[9] "He was a man who elevated the Polish nation."[10]

The two-dimensional religious-national perception of the Pope receives additional reinforcement when placed in the context of Kraków's cityscape. Kraków, among Polish people, is a very strongly ideologized city related to a concept of Polish national identity. Its "past" and its architecture (often seen as a material representation of that "past") were involved in the Christian theologization and nationalization of Kraków, especially during the nineteenth century. The Wawel Cathedral has played a leading role in these processes. The Cathedral can be analyzed as a nationally symbolic building, possessing "a kind of agency," which "provides opportunities for processes of identification" (Verkaaik 2013: 12). Up until the eighteenth century, it served as a coronation and burial place for Polish kings and queens. During the course of the nineteenth century, it was turned into a "Polish national pantheon" where national heroes, soldiers and artists began to be buried next to the old kings and queens. At this time, for Poles who opposed the division of their lands among three empires (Russia, Prussia, and Austria, 1795–1918), the space of the Wawel Cathedral was transformed into a "national temple." The Cathedral, as well as the Old Town of the city of Kraków, began to be visited by "national pilgrims." They traveled there to gain inspiration, and to educate themselves about the "Polish nation" and its "past," to learn about the "old days" when the city had been a thriving capital of a powerful European kingdom (Niedźwiedź 2015: 76).

The mythologization, nationalization and symbolization of Kraków's cityscape resulted in a combination of religious and national discourses related to the city and its buildings. Today, this proves to be one of the most productive symbolic agents attached to the common vision of Kraków among Poles. When asked about their first immediate associations with the name "Kraków," those interviewed usually recalled the historic past and national-religious heritage of the "ancient Polish royal city." A married couple in their sixties, having come to Kraków-Łagiewniki from Silesia to pray at the Divine Mercy shrine and to give thanks for a friend's successful medical operation (the operation had been performed in a Kraków's hospital named after John Paul II) explained in a way that is exemplary of many of the Polish respondents: "For us Kraków is a symbol. It is a symbol of Polish continuity. It is the historic Polish capital city. It is historical heritage of immense weight. It is what unites Poles. For us this is a very important city."[11] In my fieldwork, the Wawel Cathedral and Castle were the most frequently recalled iconic edifices of Kraków. The figure of the Pope was often enumerated next to them, as yet another "icon of Kraków" and a symbol of this "royal and papal city."[12]

Reciprocity of myths

The Pope himself was actively involved in the development of the discourse, which nationalized him (Niedźwiedź 2009). In this process, he often used the image of Kraków. Numerous speeches and sermons delivered by John Paul II in this city during his "pilgrimages to the Fatherland" emphasized the religious-national dimension of Kraków's architecture and the city's relation with the "Polish past." At the time of his first visit in 1979 to Kraków as a newly elected pope, during an open-air mass at Błonia meadow he described the city as a "synthesis of everything Polish and everything Christian." Eight years later, in 1987, standing on a high platform on the same meadow and facing the outline of the Wawel Cathedral with its slim towers marking the skyline, he welcomed a crowd of one million Polish Catholics, saying: "I am looking at Kraków. My Kraków. The city of my life. The city of our history." By formulating these words, the Pope directly inscribed himself and "his life" into Kraków's "tradition" and into the "national heritage" related to Kraków. Ethnographic material reveals that this special status of the Pope is broadly accepted among Polish people.

At this point, it may be important to emphasize the reciprocal dynamism between the symbolic potentiality of Kraków's urban space and the iconic figure of the Pope. The national dimension of the city's heritage and its scape enhance the national dimension of the Pope and the common perception of him among Poles. At the same time, his figure, being related to Kraków, confirms that Kraków is a special place for the whole nation: a "cradle of sainthood" and a source of "Polish greatness," which has not only been revealed in the remote past, but can also be witnessed today.

The relation between John Paul II and Poland's national mythology is one of the most visible aspects of the cult of the Pope in Kraków and elsewhere in the country today. As such it has been thoroughly discussed within the scholarly domain. Anthropological interpretations have pointed out the "heroic myth" of the Pope (Zowczak 1987), and its historical background dating back to nineteenth-century Polish Romantic messianism, which prophesied about the "Slavic pope" (Łozińska 1990). Jan Kubik's in-depth analysis of John Paul II's first visit to Poland in 1979 discussed the correlation between the Pope and the symbolic story of Kraków's medieval Bishop Stanislaus, who was martyred and then declared the patron saint of the Polish kingdom (Kubik 1994).

The national aspect of the popular image of the Pope appeared also in the foreground of a scholarly debate, which developed following the public mourning in April 2005. The scale of the phenomena, the emotional impact of the Pope's death on Polish society, and the spontaneity of shared behaviors are well documented in a significant body of scholarly work (Niedźwiedź 2009; Hodalska 2010; Klekot 2011; Brzozowski 2013). Anthropologists and sociologists, along with publicists, commentators and journalists, realized that such a spontaneous movement and massive communal manifestation of feelings, represented by Poles who spent hours gathering in streets, squares, soccer stadiums and other public places during the week after John Paul II's death, created an opportunity to conduct observations of nationally meaningful behaviors and spaces. The national potentiality of Kraków's cityscape was observable in a local version of mourning rituals performed in the city. Significantly, a rumor that the Pope's funeral might take place not in the Vatican, but in Kraków (at the Wawel Cathedral) was taken seriously by some Poles.

Place-making and the localized Pope

Inscribing biography into urban space

While going through my fieldwork notes from 2005 and studying transcripts from recent interviews conducted at the Kraków-Łagiewniki shrines, I realized that there is a need to go beyond the "national interpretation" when discussing the contemporary cult of John Paul II in relation to Kraków's cityscape. I wish to state that there are other parallel discourses about the Pope, which are definitely represented in the ethnographic material collected in recent years in Kraków. These discourses are of significant importance. While they do not contradict the presence and significance of the "national" dimension of the Pope's cult in Poland, they surely question its sole dominance.

Realizing this, I decided to go back to the streets of Kraków in April 2005 and look once more at the ethnographic documentation of this period that I and other researchers collected. While reading articles dedicated to the analysis of public mourning for the Pope in the streets and squares of Warsaw (Klekot 2011; Brzozowski 2014), I asked myself what might differentiate the mourning done within the cityscape of Kraków from corresponding events in the Polish capital. I was sure that the general structure of public performance and spontaneity of people's behavior had been similar, if not the same. I could not compare the emotional engagement of people living in different cities, as it was impossible to sensibly measure the intensity of people's feelings. Finally, I turned toward an analysis of sites, which, during the time of mourning, had been chosen spontaneously by people and which they had transformed into "places of remembrance."

I wish to state that the affective, personal and spiritual dimensions of experiences, as related by numerous people, associated with Kraków's "papal spaces" produce discourses which emphasize the Pope's existence as a real person, someone who was physically present within this concrete urban space. Such a potentiality of space is possible because, as expressed by one interviewee: "John Paul II was rooted here, in this city."[13] Some of Kraków's "papal spaces" enable the construction of an emotional, intimate relationship between the Pope and his present-day followers, who often relate to him while building their individual identities, overcoming their personal difficulties, or explaining their own biographies. The local dimension of the Pope experienced through Kraków's urbanscape instigates a process of privatization of the Pope. Here, the Pope comes to be seen not so much as a remote figure of a national hero uniting the Poles as a "nation," but rather as a real person, a suffering human being, whose biography might sound universal and emotionally moving to anybody. His life story is simultaneously depicted as a biography of a saint—however not a distant one, but rather a local citizen, a neighbor, a member of a family. He is seen as someone accessible and full of understanding. As someone who lived nearby, quietly developing his spiritual life. He then ended up being elected Pope, became a world-renowned figure, and, finally, was recognized as a saint.

Kraków is a city where Karol Wojtyła spent a significant part of his life. He lived there between 1938 and 1978, with only short breaks (for studies in the Vatican, and for his curate service in a small parish in the vicinity of Kraków, for example). Thus, the cityscape of Kraków is dotted with places related to various events and various stages of Wojtyła's life.

They include several buildings where he lived (as a newcomer to the city, as a student and worker, as priest, bishop and archbishop). There are places related to his studies in Polish literature at the Jagiellonian University in Kraków, to his activities as an actor performing in an underground Polish theater during the Nazi occupation, and places related to his work in a quarry and a chemical plant. Naturally, there are numerous places related to his service as a priest, then as Kraków's bishop (as of 1958) and archbishop (as of 1964). Additionally, as in Warsaw, there are places, which relate to his "pilgrimages to the Fatherland" as pope.

All these places were lived as "places of remembrance" during the week-long mourning for the Pope in 2005. A spontaneously created network of devotional and memorialized spaces and places revealed how the interaction between

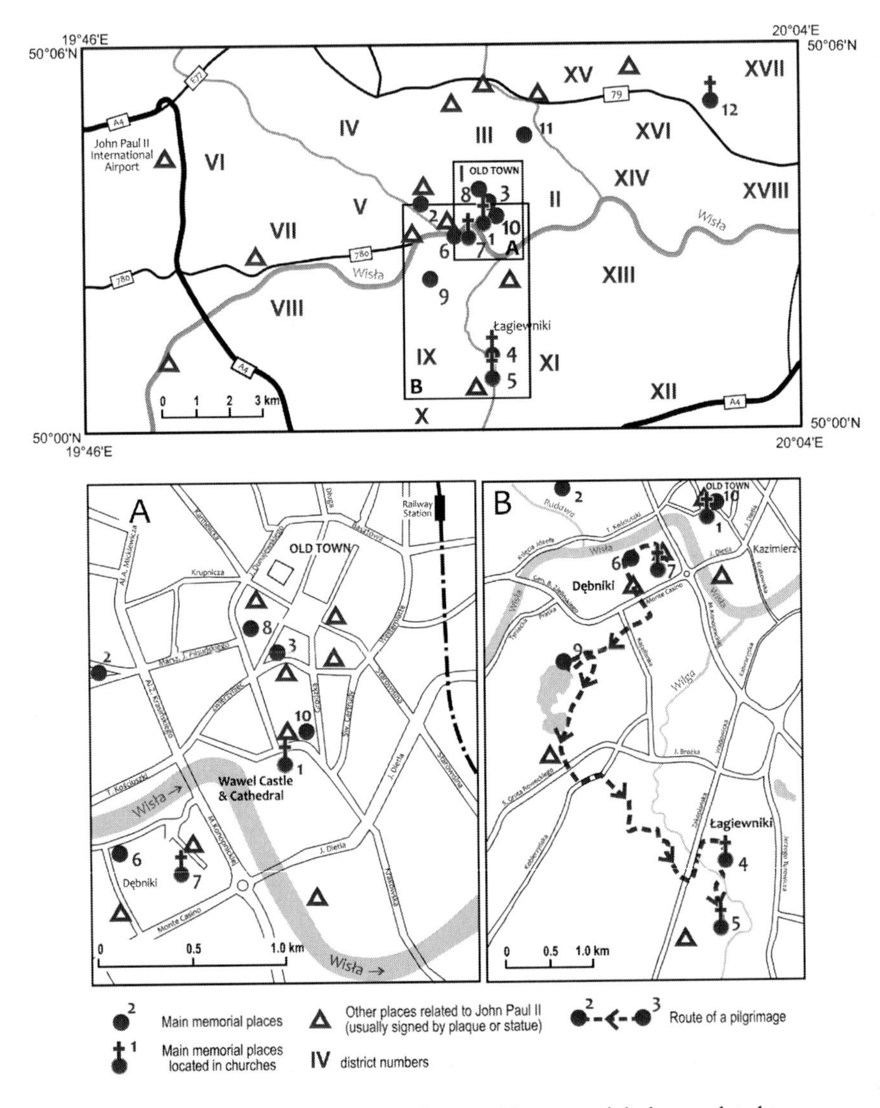

Figure 5.2 Map of Kraków and surroundings, and its memorial places related to John Paul II.

Map A. Old Town and part of Dębniki District. Map B. Route of the "Pilgrimage in the Footsteps of Karol Wojtyła the Worker." Map prepared and edited by Ewa Łupikasza.

people's religious activities, emotional reactions, and urban space in practice works. Here I will focus on the most famous and the most unusual papal memorial place in Poland: namely, the "Pope's Window." I will describe the place-making process, which turned a window on the second floor[14] of the Bishops' Palace into "the most famous window in Kraków" (Skowrońska 2008: 5), known today all over the country as the "Pope's Window."

The "Pope's Window"

In 2005, Easter was celebrated on March 27; many people remember this Sunday as the last public appearance of John Paul II. Three days after the Easter holiday, I was working in my office at the Jagiellonian University, located just two minutes' walk from the Bishops' Palace. In the afternoon, one of my colleagues, entering my department, announced with a strained voice that he had just seen people gathering in front of the Bishops' Palace. This, as he put it, meant that "something is going to happen." That day I left my office late in the evening and decided to pass the Bishops' Palace to see what was going on. I saw a group of several dozens of people, praying in silence and intensely looking toward a central window, located on the second floor of the Bishops' Palace, directly above the main entrance. As a native of Kraków, I knew the significance of this window in relation to John Paul II. First of all, it was one of the windows of the private apartments of Kraków's archbishops. Further, this window also served as an informal meeting point for the Pope and the inhabitants of Kraków when Wojtyła, as pope, visited the city during his "pilgrimages to the Fatherland."

When John Paul II visited Kraków during his first papal trip in June 1979, in the evening, after the official part of the visit, crowds of Kraków residents spontaneously gathered in front of the palace to praise the Pope, who, just a few months earlier, had been their Archbishop. They were singing, shouting and calling the Pope's name. Finally, the window on the second floor opened and John Paul II appeared. Using a microphone that had been quickly set up, he started joking with the gathered Kraków residents recalling his days as a regular citizen of the city (asking about actual local places and jokingly complaining about changes in his life when he had moved to Vatican, for example). He went on to sing with the people, and finally asked them to join him in prayer. In this way, the public city area was informally turned into a space for an open-air religious event, bringing up a public gathering that emphasized a private and almost familial dimension of the Pope for the local people. The evening ended late at night with his papal blessing. This local ritual, initiated during that spontaneous first meeting in 1979, was repeated each time the Pope stayed overnight in Kraków. During his pontificate, there were 16 informal evenings in front of the Bishops' Palace (Jakubczyk and Tekieli 2005: 103). These meetings were never formally arranged or officially announced. However, whenever the Pope was staying in the Bishops' Palace in Kraków, the enthusiastic crowd would gather, sing and shout, and wait as long as necessary to have an informal meeting with "their Pope."

After each visit of John Paul II to Kraków, the window in the Bishops' Palace would go back to "normal." Interestingly, relying on my own memories and of those close to me, I am not able to recall anybody in Kraków using the term "the Pope's Window" before April 2005. For us it was just a regular spot. We knew that it was associated with Wojtyła, and it was obvious that during his papal visits the place was used for informal mass audiences with the Pope. Still, nothing special was associated with the Bishops' Palace and its windows on a day-to-day basis. Soon after the first papal visit to town, a monument of the Pope was erected in the courtyard of the Bishops' Palace. This courtyard was not visible from the street and not very well known to the wider public. Thus, I was amazed and fascinated when at the end of March 2005, during my evening walk home from my office, I encountered a small group of people silently gathering in front of the Bishops' Palace. They were gazing at the dark window on the second floor. Soon afterward, this informal public "papal space" would undergo a dramatic transformation.

Following the news from the Vatican, which, day by day, and, later, hour by hour, informed the world about John Paul II's deteriorating condition, the gathering in front of the Bishops' Palace grew in numbers. Loud prayers and songs began to be heard. Someone brought the first candles and flowers and placed them in front of the palace. As early as on Friday, April 1, the rapidly growing number of people keeping vigil in front of the building forced the City Council to close off the entire street to traffic. Cars and public transportation, including a very popular tramline, were rerouted. (The detour lasted more than a week. The area around the Bishops' Palace was only cleaned up after the Pope's funeral. Garbage collectors worked for two days collecting leftover candles and flowers.) In the afternoon, the crowd became so densely packed that the City Council decided to put up metal barriers to regulate movement and to keep the crowd a few meters away from the walls of the Bishops' Palace. These walls were already covered with posters, stickers, and banners expressing love and gratitude to the Pope. The flames from candles placed on the pavement and on the outer sills of first-floor windows covered the walls with soot. Soon candles occupied the entire area in front of the palace, spreading out along the street, and appearing as well in various spots around a nearby park, and a small square on the opposite side of the street.

The next day was Saturday, April 2, and literally thousands of people surrounded the Palace starting in the early morning hours. (Some had even kept an overnight vigil.) On this day, a window on the second floor was opened and a wooden crucifix with a red stole (the symbol of the death and resurrection of Christ) appeared in the middle of it, decorated with white and yellow flowers (the Vatican's colors).

More and more candles, flowers, notes and handwritten letters to the Pope appeared in various locations in the vicinity of the palace. Fans of Kraków's soccer teams began to hang their team scarves and t-shirts on the metal barriers, which now separated people from the sea of burning candles. When, late in the evening, just minutes before 10 p.m., the news of the Pope's death reached the crowd, people stood in silence. Then they began to pray and to intone a traditional Easter song about resurrection. In nearby churches as well as at the Wawel Cathedral, masses for the soul of the departed Pope were held at midnight. Throughout the night, people gathered in various locations, turning these spaces into spontaneously arranged "places of remembrance." For the whole week, named the "Week of Vigil," the "Pope's Window," as it began to be called by people and by the media, remained a central spot for mourners gathering in Kraków.

Figure 5.3 The "Pope's Window," a few days after the death of John Paul II, April 7, 2005. Photo by Anna Niedźwiedź.

The life of the "Pope's Window"

The "Week of Vigil" lasted until the funeral of the Pope (April 8). The pattern of behavior and feeling of *communitas* was experienced similarly in various cities and places in Poland (Klekot 2011: 198). I recall the story about the "Pope's Window" here to show that, in the case of Kraków, the local perception of the Pope as an inhabitant of the city was of great importance in the place-making process. While numerous spontaneous "places of remembrance" disappeared soon after the intense period of mourning, and others were strongly institutionalized by the Catholic Church, which introduced official ways of celebrating the Pope (through annually organized open-air masses, in the center of Warsaw, for example; see Brzozowski 2013: 252), the Kraków window represents an interesting combination of both: grass-roots and institutionalized activities related to its contemporary *life*.

Since April 2, 2005, Kraków's archbishops who reside in the Palace and use its living quarters, have taken care of the Pope's Window and arranged its appearance. Soon after the funeral, a huge portrait of the Pope was placed in the frame of the window, covering it completely. Interestingly, over the course of the ensuing years, the portrait has been changed frequently. I remember seeing different versions of "the Pope" exhibited in the window. For a substantial period, there was a portrait depicting John Paul II at a very advanced age, with a walking stick in his hand, with clear signs of suffering depicted on his face. Since the Pope's canonization, the official portrait of St. John Paul II has been exhibited in the window.

Institutionalized events organized by the Church in front of the Pope's Window include celebrations of the anniversary of the Pope's election or events related to his canonization. Church-organized events usually include some open-air concerts and prayer meetings. However, most intriguing is the grass-roots life of the window, which has been developing on its own. As my office is located nearby, and I usually pass the Bishops' Palace on my way to work, I have had the opportunity to note various informal activities in front of the window. Almost every day, it is possible to see flowers or new candles appearing on the other side of the street, opposite the palace, the best place from which to view the window. Interestingly, the annual cycle of anniversaries and feasts observed in front of the window in the form of spontaneous activities is much richer than the cycle of institutional events organized by the Church. Religious feasts, which in Poland are usually celebrated within the family circle, trigger some grass-roots activities around the window. At Christmas and Easter, more candles and flowers are deposited in front of the Pope's Window. Sometimes, seasonal Christmas or Easter decorations appear next to them. All Saints' and All Souls' Days turn the place into a "memorial ground," where people come to burn candles (as they do in cemeteries at the graves of their loved ones on these days). Wojtyła's name day and birthday are also marked by similar activities.

On any normal day, passers-by respond to the presence of the Pope's Window. Some give it a prolonged look, many mark it with at least a glance. It is also very interesting to observe the behavior of people aboard the tram which passes the "Pope's Window." It is not unusual to see somebody making the sign of the cross, while looking at the portrait of the Pope in the window's frame. Various, and not only religiously oriented, visitors come to the vicinity of the Bishops' Palace in order to see the window. Many tourist groups from Poland and from abroad are led here by city tour guides. Individual tourists also come, as the window has appeared in guidebooks, described as a new "must-see" tourist attraction in Kraków. Responding to growing interest, a nearby newspaper stand has begun to offer a selection of candles, papal memorabilia and postcards featuring the "Pope's Window." Some people evidently come to pray in front of the window

(I myself have seen several groups praying there). Many take pictures of the Pope's Window, eagerly posing for pictures with the window in the background, or taking selfies. Some have begun to treat the window as a sacred relic. During the 2006 renovation of all the windows in the Bishops' Palace, many were concerned about the renovation or possible replacement of the "original Pope's Window." Church officials publicly assured the city's inhabitants that, following the renovation, the "Pope's Window's" original frame would be restored, but covered with additional protective glass, as "no one in Kraków would dare to replace the old frame" (Skowrońska 2008: 5).

"One of us"

Kraków's spaces seem to provide double frames for the figure of the Polish Pope: it places him not only within Poland's national heritage but also emphasizes his existence as a real person. Personal, affective and spiritual relations between people and "their Pope," whose biography is inscribed into urban space ("Maybe we are even walking in his exact footsteps"[15]), accompany the national discourse, sometimes undermining its dominance, while providing a different, private and local, story about John Paul II.

The Pope's Window itself recalls the idea of "home" and of something very human and familial. Memories of informal meetings with the Pope in front of the window describe him first of all as a local citizen of Kraków: "one of us." His relationship with the city is expressed in terms of home:

> He [the Pope] felt a special affection for this place [Kraków] . . . You know, it is like home. One leaves home. And then it does not matter what part of the world one lives in; there is always this attraction to home.[16]

Ethnographic material reveals very affectionate relationships with the Pope. Many interlocutors relate to him as to a family member, especially when recalling his death. To my surprise, even in interviews which were carried out nine years after the Pope's funeral, numerous interlocutors compared his death to the loss of a loved one: "It was as if someone very close, someone very dear to one's heart, was dying."[17] "I cried . . . as if it had been my Grandpa."[18] "It was as if . . . as if my husband had died . . . as if the nearest and dearest person had died."[19]

The personalization, privatization and localization of the Pope are also evident in the case of another Krakówian practice: the "Pilgrimage in the Footsteps of Karol Wojtyła the Worker." This is an annual pilgrimage on foot. It is organized by priests and parishioners from a parish in the Dębniki district of Kraków, where Karol Wojtyła lived when he and his father first moved into the city in 1938. He lived there during the Nazi occupation and this is where his turn toward spiritual life and a priestly vocation occurred. (In 1942, Wojtyła enrolled in an underground seminary.) During the Nazi occupation, universities and higher educational institutions in Kraków were closed; individuals aged 18 and older were supposed to be employed, mostly in industrial production (Chwalba 2002: 382). Thus, Wojtyła (while studying at an underground university) started working in a quarry and in a chemical plant located in the southern suburbs of Kraków-Łagiewniki. The "Pilgrimage in the Footsteps of Karol Wojtyła the Worker" focuses on this part of Wojtyła's biography.

The pilgrimage is designed by its organizers as a walk covering the daily route which Wojtyła followed from his home in Dębniki to the quarry and to the chemical plant.

There are a few "stations" during this route, which recall places and events related to the personal and spiritual biography of Wojtyła. Personal suffering (solitude, loss of his parents and brother) and spiritual development are elements, which are meditated upon during the pilgrimage and presented in the context of Christian doctrine. Stations of the Cross, performed and prayed inside the picturesque old quarry, are the most moving part of the pilgrimage and are deeply and emotionally experienced by its participants. Interestingly, even though the pilgrimage recalls the years of Nazi occupation, the suffering of Wojtyła is hardly ever described within the context of national suffering. The pilgrimage focuses exclusively on the personal and spiritual life of Wojtyła: it depicts him in terms, which enable pilgrims to identify with his experiences and to gain strength and hope from his life story. Wojtyła's time as a worker and the emphasis put on his very poor and modest life during this period have a contemporary ring for many of today's pilgrims. They themselves often encounter difficulties related to work (or to unemployment), as well as to their economic situation. The litany used at the end of the 2013 pilgrimage named John Paul II as one who is "supporting the unemployed, concerned about the homeless, visiting prisoners, supporting the weak ones, teaching everybody solidarity."

The "World Center" and the globalized Pope

Two shrines in Kraków-Łagiewniki

Since 2012, the "Pilgrimage in the Footsteps of Karol Wojtyła the Worker" has found its final stop in the newly built shrine dedicated to John Paul II. This shrine was built on the suburban site of the former chemical plant where Wojtyła worked during World War II. The John Paul II shrine has been constructed in the immediate neighborhood of the Basilica of Divine Mercy, which was erected in 2002 and attracts two million pilgrims annually (Mróz 2008: 54). It used to serve as the final stop for the pilgrimage in the years preceding the construction of the John Paul II shrine. Still, it is almost impossible to avoid passing through the area belonging to the Divine Mercy shrine in order to reach the John Paul II shrine.

Both shrines constitute a large religious complex dedicated to two new, modern and intertwined devotions: the devotion to the Divine Mercy and its Polish "apostle," Saint Faustina Kowalska, a sister in the Congregation of Sisters of Our Lady of Mercy, who is believed to have prophesied about the "Polish Pope," and devotion to John Paul II—being the former propagator of the Divine Mercy itself and instigator of the canonization of Sister Faustina Kowalska.

Figure 5.4 "The Pilgrimage in the Footsteps of Karol Wojtyła the Worker" passes by the Divine Mercy Basilica in Kraków-Łagiewniki, October 2013. Photo by Anna Niedźwiedź.

Both devotions are strongly associated by Polish people with the cityscape of Kraków (as both Faustina Kowalska and Karol Wojtyła spent significant parts of their lives in this city).

The construction of these two large modern churches has changed the spiritual landscape of the city of Kraków. It has created a new religious center with modern Christian architecture and a big green area surrounding it, located outside the historic and heritagized Old Town. The green area in Kraków-Łagiewniki is used for open-air masses during bigger religious feasts. Numerous inhabitants from the suburban district also treat it as a local park on a daily basis; it is a safe area to sit and relax on benches or spend time strolling with children, sometimes combining a walk with a brief visit to one of the shrines for a prayer. The relation between the two churches (the Divine Mercy Basilica and the shrine of John Paul II) bears a symbolic significance. Many interlocutors emphasized that there could have been no Divine Mercy shrine without John Paul II. They were citing his personal relationship with Divine Mercy, which started during the years of Nazi occupation and lasted until the end of his life:

> John Paul II is from Kraków, and it is because of him that the basilica of Divine Mercy appeared and the cult of Divine Mercy benefited so much!"[20]
> "There is a link between the two sanctuaries. This link is John Paul II, who,

> as Pope, established the feast of Divine Mercy. Apart from this, John Paul II often came here [meaning the old cloister in Łagiewniki]. He prayed here. He knew Łagiewniki. Kraków was a place very close to him and I think this is a very good place for his shrine. And he personally venerated Divine Mercy, so the location of his own shrine next to the Divine Mercy shrine is fully justified.[21]

The shrine of Divine Mercy originates in the figure of Faustina Kowalska, a visionary nun, who died in Kraków-Łagiewniki in 1938 at the age of 33. During her life, she reported numerous apparitions of Christ, which she described in her spiritual diary. The idea was to promote the devotion of Divine Mercy among all people, to create a place dedicated to this cult, and to show a holy image depicting Christ as he appeared in her visions (Garnett and Harris 2013). After the death of Sister Faustina, the cult of Divine Mercy developed rather slowly and mostly on a grass-roots level (Czaczkowska 2012: 335). In the Łagiewniki suburb of Kraków, where Faustina spent her final years and where she was buried, the cult was known among local people who visited the small cloister's chapel and attended mass there. During the Nazi occupation, the cult gained some new followers. In 1942, a Kraków artist, Adolf Hyła, painted an image of Christ based on Faustina's vision and donated it to the chapel as a votive offering for saving his family during the turmoil of the Nazi occupation. Later, this image was to become the most popular representation of the Divine Mercy, a devotion that recently has become popular all over the world.

However, originally the cult as well as the image itself stirred some theological controversy (Czaczkowska 2012: 336–56). Until June 1978, the cult of Divine Mercy, in the form promoted by Sister Faustina, was officially banned by the Catholic authorities. It was the sustained efforts of Karol Wojtyła (first as Kraków's bishop, later as pope), which led to the acceptance of the cult and later to its skyrocketing popularity (ibid.: 356–7). His efforts brought about the canonization of Sister Faustina (2000) and the erection of the huge Divine Mercy basilica in Kraków-Łagiewniki (2002), which is called the "World Center of Divine Mercy." Thus, the recent erection of a new monumental church dedicated to St. John Paul II immediately next to the Divine Mercy shrine emphasizes the relationship between these two figures and "their places in Kraków": the site where Faustina lived and died, and the site of the chemical plant where Wojtyła once worked and where he learned about the Divine Mercy devotion emerging in the neighboring cloister.[22]

Transcending the national discourse

The Divine Mercy shrine that, since the fall of Communism in Poland, has been transformed from a local site bearing the label of "unofficial" into a "world center," is sometimes seen as a new shrine for a new time or world to come. It supposedly proposes the ideal of "open" or "worldly Catholicism" breaking with the Polish national-Catholic discourse. It is said to promote a global vision of religion and spirituality, which embraces all people from all different nations, races and continents. Polish pilgrims encountered at the shrine often invoke this global perspective, as in the case of a man in his late fifties:

> I am personally fond of the Divine Mercy Hour. This is [the hour] when we connect with the whole world. There is the praying of the rosary in different languages [the Divine Mercy rosary is said in the Divine Mercy shrine and in the John Paul II shrine every day at 3 p.m. to recall the hour associated with Christ's death, thus 3 p.m. is called "the Divine Mercy Hour"]. And it makes me think that it is not only us praying here, but that the whole globe is praying with us. Because it is in different languages, in English, and others . . . Usually it depends which pilgrimages arrive on a given day. One-tenth of the chaplet is said in their language, and so on. It's a very beautiful prayer. It's the "Hour of Mercy" for the whole world.[23]

The Pope, as associated with the "World Center of Divine Mercy," is also depicted by numerous interviewees in slightly more "globalized" terms, transcending the "national discourse." Some pilgrims link John Paul II to the prophecy from Sister Faustina's diary in which she described "the spark that will come from Poland to prepare the world" (for the "final coming" of Christ) (Kowalska 1987: 612). Interestingly, in collected interviews, the apocalyptical dimension of this prophecy (clearly present in Faustina's diary) seems to be reformulated. People explained that in their views the "spark" means the Pope or Divine Mercy itself, "which come from here to the whole world . . . This is that Center [meaning the Divine Mercy shrine] and the Pope who in this place . . . entrusted the whole world to Divine Mercy."[24] That is why, when visiting the shrine, people can experience "this spark, which started glowing here and is kindling the whole world."[25] In these explanations, John Paul II appears as a fulfillment of Faustina's prophecy. Being a "pope from Kraków," he popularized the local cult, made it global and stopped or postponed the "final coming" of Christ, helping to reveal God's mercy to the people all over the world.

Reflecting on the meaning of the spaces they experience in Łagiewniki, visitors often describe John Paul II not only as a Polish, but also as a world leader. An older woman, describing the various chapels located in the Divine Mercy basilica, praised their national variety (the Slovakian, German, Italian, Hungarian and Ukrainian chapels, for example), simultaneously emphasizing the transnational unity of the Catholic Church, which includes the global institution of the papacy.[26] John Paul II himself is mentioned as someone whose story is universal enough to attract pilgrims and devotees from all over the globe. Polish pilgrims recalled foreigners they had met in Łagiewniki's shrines and the foreign languages they had heard there.

Wojtyła is also mentioned as someone who initiated interreligious and interdenominational dialogue. He was also frequently mentioned as the one who initiated the Catholic World Youth Day. As the next World Youth Day was scheduled to take place in Kraków in summer 2016, those interviewed in 2014 expressed their opinions about this upcoming event. Some emphasized that such a global meeting is a chance to promote Kraków to pilgrims and tourists from abroad. It can also be a good opportunity to share various experiences and meet people from different cultures. A woman in her early thirties, interviewed in front of the Divine Mercy shrine, enthusiastically described the idea of ecumenical meetings and interreligious dialogue. She emphasized that John Paul II propagated these ideas in the media, which is why Łagiewniki, as a place that memorializes him, is especially suitable for the fostering of these ideas. When reflecting on the planned World Youth Day, she pointed out that this event would surely attract not only Catholics, and added: "I think it is not only a matter of deepening or widening one's faith, but also a matter of personal meetings and direct encounters with different cultures, which also open our minds to other religions."[27]

Conclusions

During the Communist period, references to the Polish Pope generated a strongly nationalized discourse, making him a national liberating hero and the savior of the nation. The fall of Communist Poland (in 1989) was followed by rapid political, social and economic transformations, which led to the creation of a democratic system and an open-market economy. The lifestyles of Poles have changed tremendously. On one hand, they have praised access to new possibilities and the liberation of the flow of people and ideas. On the other hand, the transformation has caused a communal feeling of discontinuity and social instability. Thus, the figure of the nationalized Pope was treated by many as a mental bridge and a stable point of reference. As highlighted by Brzozowski, John Paul II's "pontificate lasted for 27 years and managed to connect a number of generations of Poles," who started to treat him as "an embodied evocation of extended, historical temporality" (2014: 89). Continuation of the national discourse has thus been one of the ways to make sense of a changing reality. This national discourse is still clearly visible in the ways in which John Paul II is inscribed and lived within Kraków's cityscape today. This combines historical memories of the "Polish pantheon" with the more contemporary image of the "Pope's city," making John Paul II a part of the "national heritage."

The figure of the Pope framed within Kraków's urban and suburban spaces also offers other aspects. He is seen as a local person: a real inhabitant of the city, as "one of us." Through the lens of Kraków's part of his biography, Karol Wojtyła easily gets privatized by Poles today. The privatization of the Pope focuses on his personal suffering and spiritual development. His time as a worker enables many people to relate to him on a very individual level. In this respect, Kraków reveals its spatial potentiality, embracing grass-roots sacred places and generating new rituals, which emphasize a less official (and less nationalized) part of Wojtyła's biography (such as the "Pilgrimage in the Footsteps of Karol Wojtyła the Worker").

Finally, within the period of more than 25 years that have passed since the fall of Communism in the country, Polish Catholics have realized more clearly that Catholicism is not only a Polish matter. (The national dimension of Polish

Catholicism played a crucial role in resisting the Communist state and opposing anti-religious propaganda in Communist Poland.) Catholicism has come to be seen as, in fact, a "religion of the world." Within the last two decades, the mobility of Poles has grown rapidly. Polish people emigrate in great numbers and encounter people from different cultures and backgrounds. The "global Pope" and the "global cult" of the Divine Mercy, based within the newly built shrines in Kraków's suburban cityscape provide sacred ground and a symbolic space. This space enables many Poles to locate themselves and their new devotions and spiritualities within a newly-opened global context. The multivocal, dynamic, produced and productive cityscape of Kraków enables contemporary Polish Catholics to locate themselves within different contexts applicable to their lives. Answering various needs, the generative space of Kraków is being *lived* by many Poles as the "Pope's city" within nationalized, localized and globalized discourses.

Notes

1 The written transcripts of 200 interviews conducted in Kraków-Łagiewniki in August and September of 2013 and 2014 are stored in the Archive of the Institute of Ethnology and Cultural Anthropology of the Jagiellonian University in Kraków. I also use 30 transcripts of interviews carried out by myself and by students attending my ethnographic fieldwork seminar in 2005 and 2008. When quoting field research material I provide a numeric code representing a particular interview, the year, and, if available, the exact date. Here I would like to thank undergraduate students involved in the ethnographic fieldwork in Kraków-Łagiewniki in 2013: Aleksandra Kopeć, Agata Kuliczkowska, Natalia Kuta, Jakub Liszewski, Ewelina Luberda, Dagmara Masłowska, Grzegorz Mołda, Marek Niewadzi, Emilia Pyrz and Patryk Trela; and in 2014: Karina Kiwior, Aleksandra Kopeć, Maciej Krupa, Jakub Liszewski, Paulina Maciejasz, Natalia Migdał, Karolina Molenda, Marta Skrzekut, Jaonna Skrzydlewska, Agnieszka Ziaja. Additionally, I would like to thank Alicja Soćko-Mucha, PhD student from my department, who assisted me in supervising students during the 2013 fieldwork.

2 015_2013, September 6.

3 120_2014, August 30.

4 060_2013, September 13.

5 091_2013, August 31.

6 003_2005, November 4.

7 003_2013, September.

8 018_2013, September 7.

9 014_2013, September 3.

10 120_2014, August 30.

11 133_2014, August 30.

12 133_2014, August 30.

13 155_2014, August 30.

14 First floor in British terminology; second floor in American terminology.

15 062_2013, September 8.

16 001_2008, September 8.

17 114_2014, September 5.

18 101_2014, August 31.

19 141_2014, August 29.

20 019_2013, September 8.

21 014_2013, September 3.
22 It is important to add that a significant number of our interlocutors emphasized the link between Divine Mercy and John Paul II, stating that it would be perfectly suitable to create a place dedicated to the cult of the Pope directly within the space of the existing Divine Mercy shrine. Thus, some expressed criticism that two huge separate churches located next to each other might cause some unnecessary "competition" between those two, strongly correlated devotions, since people when arriving in Kraków-Łagiewniki will have to decide in which church they want to spend more time, take part in a holy mass, etc.
23 200_2014, September 8.
24 190_2014, September 9.
25 001_2013, August 31.
26 180_2014, September 7.
27 110_2014, September 8.

References

Bianchini, Franco (2006), Introduction. European Urban Mindscapes: Concepts, Cultural Representations and Policy Applications, in Godela Weiss-Sussex and Franco Bianchini (eds), *Urban Mindscapes of Europe*. Amsterdam and New York: Rodopi.

Bielo, James S. (2013), Urban Christianities: Place-Making in Late Modernity, *Religion*, 43: 301–11.

Brzozowski, Grzegorz (2013), Spatiality and the Performance of Belief: The Public Square and Collective Mourning for John Paul II, *Journal of Contemporary Religion*, 28: 241–57.

Brzozowski, Grzegorz (2014), Clashing Temporalities of Public Mourning: Warsaw after the Death of Pope John Paul II, *The Drama Review*, 58: 84–96.

Chwalba, Andrzej (2002), *Dzieje Krakowa. Kraków w latach 1939–1945*. Kraków: Wydawnictwo Literackie.

Czaczkowska, Ewa K. (2012), *Siostra Faustyna. Biografia świętej*. Kraków: Wydawnictwo Znak.

Duff, Cameron (2010), On the role of affect and practice in the production of place, *Environment and Planning D: Society and Space*, 28: 881–95.

Eade, John, and Christopher Mele (2002), Introduction. Understanding the City, in John Eade and Christopher Mele (eds), *Understanding the City: Contemporary and Future Perspectives*. Oxford and Malden, MA: Blackwell.

Eade, John and Mario Katić (2014), Introduction: Crossing the Borders, in John Eade and Mario Katić (eds), *Pilgrimage, Politics and Place-Making in Eastern Europe: Crossing the Borders*. Farnham and Burlington, VT: Ashgate.

Garnett, Jane and Alana Harris (2013), Canvassing the Faithful: Image, Agency and the Lived Religiosity of Devotion to the Divine Mercy, in Giuseppe Giordan and Linda Woodhead (eds), *Prayer in Religion and Spirituality*. Leiden and Boston, MA: Brill.

Hodalska, Magda (2010), *Śmierć Papieża, narodziny mitu*. Kraków: Wydawnictwo Uniwersytetu Jagiellońskiego.

Jakubczyk, Michal and Roman Tekieli (2005), *Kraków, miasto mojego życia. Przewodnik śladami Jana Pawła II*. Kraków: Katolickie Centrum Kultury & Wydawnictwo M.

Klekot, Ewa (2011), Mourning the Polish Pope in Polish Cities, in Peter Jan Margry and Cristina Sánchez-Carretero, *Grassroots Memorials: The Politics of Memorializing Traumatic Death*. New York and Oxford: Berghahn.

Kowalska, Maria F. (1987), *Divine Mercy in My Soul: Diary of Saint Maria Faustina Kowalska*. Kraków: *Misericordia* Publications.

Kubik, Jan (1994), *The Power of Symbols against the Symbols of Power: The Rise of Solidarity and the Fall of State Socialism in Poland*. University Park: Pennsylvania State University Press.

Łozińska, Kinga (1990), "Więc oto idzie papież słowiański . . . " – romantyczne tło mitycznego wizerunku Jana Pawła II, *Prace Etnograficzne*, 27: 37–46.

Mróz, Franciszek (2008), Geneza i funkcjonowanie sanktuariów Bożego Miłosierdzia w Polsce, *Peregrinus Cracoviensis*, 19: 47–72.

Niedźwiedź, Anna (2009), Mythical Vision of the City: Kraków as the "Pope's City," *Anthropology of East Europe Review*, 27: 208–26.

Niedźwiedź, Anna (2015), Old and New Paths of Polish Pilgrimages, in Dionigi Albera and John Eade (eds), *International Perspectives on Pilgrimage Studies: Itineraries, Gaps and Obstacles*. New York: Routledge.

Orsi, Robert A. (1999), Introduction. Crossing the City Line, in Robert A. Orsi (ed.), *Gods of the City: Religion and the American Urban Landscape*. Bloomington and Indianapolis: Indiana University Press.

Skowrońska, Małgorzata (2008), *Spacerownik papieski. Specjalny dodatek do Gazety Wyborczej*, Kraków, October 16 (newspaper addition).

Tweed, Thomas A. (2011), Space, *Material Religion*, 7: 116–23.

Verkaaik, Oskar (2013), Religious Architecture. Anthropological Perspectives, in Oskar Verkaaik (ed.), *Religious Architecture: Anthropological Perspectives*. Amsterdam: Amsterdam University Press.

Zowczak, Magdalena (1987), Jan Paweł II – narodziny legendy, *Literatura Ludowa*, 31: 3–12.

6 Religious mediations in a dense cityscape

Experiences of Catholic and Buddhist spaces in Hong Kong

Mariske Westendorp

On a rainy Saturday morning in June 2014, around 9:00 a.m., I entered the "TML" building in Tsuen Wan West, a district in the New Territories of Hong Kong, a region located on the southern coast of China. The TML tower, a 30-story high, new and modern-looking industrial building, was completed in 2013.[1] I walked through the imposing entrance hall, with its high ceiling, leather chairs and large windows to the elevator. On my way, I saw a few people, wearing yellow vests featuring a picture of a lotus flower on the front pocket, giving directions to visitors like me. I took the elevator up to the sixteenth floor. When the doors opened, I saw several people walking around, busily arranging everything for the special occasion that was about to start in half an hour. They seemed nervous. The event had been convened to celebrate the opening of a new Buddhist mindfulness center in one of the offices of this industrial building.

I entered the office space, a room of approximately 100 square meters. In a little shop located on the right side of the office, a woman was selling Buddhist statues, beads, books, cushions and other objects. Behind the shop, in the middle of an otherwise empty space, dozens of people were already standing or sitting on the floor, some in the lotus position. Looking straight ahead, I could see a special chair for the Buddhist master and an altar with some Buddhist statues on it. In front of the windows at the back of the square office stood a long table; invited clergy would sit behind this table during the ceremony. I sat cross-legged between the other people present, waiting for the ceremony to begin. I greeted people who were familiar to me as they arrived. Among them were a few familiar faces, people I used to see at a nearby, historical Buddhist temple (built in 1933). I spent many afternoons at the temple during my period of research.[2] Until recently, the Malaysian monk who would run this mindfulness center, had been the temple abbot. The reason why these people had assembled in this office space was obviously not the venue: no one would opt in favor of a small office in an industrial building rather than visit a Buddhist temple. The people had instead come to hear the Malaysian monk's *dharma* teachings.

At 9:30 a.m., the ceremony started with Pali chanting and offerings being made to the Buddha. Then, a Buddhist master (the teacher of the Malaysian monk) took his seat in the large wooden chair in front of the crowd, which had swelled to over 150 people. He started his *dharma* sharing. While listening to

Figure 6.1 Listening to Buddhist *dharma* teachings in an industrial office. Photo by Mariske Westendorp.

the teachings, a particular comment stuck in my mind. According to the Buddhist master, "in Pali, a temple is called a *Vihara*. This word has two meanings: a physical building, or a place of refuge for the mind."[3] According to the master, a temple's function is to be a place where the wise can teach the laity. I found these words intriguing. As I sat on the floor of this office space decorated as a Buddhist center, the master's words evoked in me the understanding that a temple does not necessarily have to be a specific physical building in order to fulfill its function. It follows from this that a temple is a place that can transcend boundaries such as those that bind a specific place, and separate what is perceived as the "sacred" from the "secular." This transcendental character of a Buddhist temple has consequences for the ways in which people experience such religious buildings, and ties into people's notions of the relationship between built religious structures and religious practice, and, by extension, appreciation of the geographical location and architectural design of a religious building.

The transcendental character of religious buildings likewise came to the fore in conversations I had with visitors to the nearby Annunciation Church, a Catholic church situated in Tsuen Wan approximately one kilometer north of the TML tower. This church, which is one of the newest church buildings in Hong Kong, was built in 1993. It is one of 97 places of Catholic worship in Hong Kong divided over 51 parishes (Catholic Truth Society 2013: 666). Unlike the space that housed the Buddhist mindfulness center, this religious building is highly visible in the landscape of this part of Tsuen Wan. Situated on a crossroad next to a large, newly renovated hospital, it has a large, distinctive white cross on top. A round building, it is located next to a bus stop named after the church.

The church was designed by an Italian missionary priest, who wanted to build a specifically Italian church with Italian features. In an interview in early 2013, the priest said: "I built an Italian church. The Annunciation Church is an explosion of color, with the mosaic at the back and the stained glass windows." Whereas for the priest, these Italian features rendered the Annunciation Church unique, I will suggest that for visitors to the building, the church is similar to others as it is representative of a global network.

I will now explore the relationships between religious space and religious practice in Hong Kong. I will show how the particularity of life in Hong Kong, specifically the dense character of the built environment, inscribes itself in meanings attached to specific religious spaces. I will do this by analyzing the narratives of visitors to the Buddhist mindfulness center alluded to above, and compare these narratives to those of Catholic practitioners visiting the Annunciation Church. By comparing these two religious buildings, I will be able to indicate how Buddhist and Catholic individuals experience these spaces, and how their experiences are inscribed in the meanings they attach to Buddhist temples and Catholic churches in general. This will allow me to show that in Hong Kong, specific religious buildings are not the most important medium for Hong Kong religious practitioners; nor are they a critical precondition for practice. Rather, what is of value is what these buildings represent.

The anthropological insights I provide in this chapter are intriguing for three reasons. First, they shed light on the concept of the urbanesque, a term that the editors of this book apply to the specificity and diversity of small and mid-sized town life as well as suburbia and the rural-urban periphery and exurbs. The value of the term lies in the fact that it includes interactions between the cityscape and the wider urbanesque environment. The term thus alerts scholars to the mobility between these two environments; for example, by researching how and why in any given city religious centers are pushed to the outskirts of the city center. However, the term fails to capture the mobility that takes place within the cityscape. In this chapter, I will demonstrate how in the unique environment of Hong Kong, mobility does not necessarily take place between city center and urbanesque environment: it mainly occurs within the cityscape. Due to rapid population increase, Hong Kong's cityscape has expanded into a wider urbanesque environment, slowly transforming all environments into cityscapes while at the same time becoming denser. With regards to this process, I will now refer to Hong Kong as a "region," a term that incorporates both its cityscape and its wider, ever-decreasing urbanesque environment.

Tsuen Wan (the district in which the Annunciation Church and the Buddhist center for mindfulness are situated) is a good example of how a cityscape expands into the urbanesque, transforming it into a cityscape and becoming increasingly densely settled. One of Hong Kong's 18 districts, Tsuen Wan is located in the west of the New Territories. The latter, which are part of Hong Kong, were developed last and for this reason are less densely populated than the other two areas (Hong Kong Island and the Kowloon Peninsula). The Hong Kong Tourism Board describes the New Territories as follows:

> Despite its image as a forest of towering skyscrapers, nearly half of Hong Kong is actually given over to soaring mountains laced by ancient footpaths, long sweeping beaches, working vegetable farms and treasured Taoist and Buddhist temples. This is the *rural land* between urban Kowloon and Mainland China, long known as the New Territories, which is linked to the big city by modern highways.[4]

However, despite this "rural" image, large areas of the New Territories have become part of Hong Kong's cityscape, Tsuen Wan being a case in point. According to Lai and Dwyer, Tsuen Wan was one of the first districts "of industrial development in Hong Kong outside the urban areas in the immediate postwar period" (1964: 153). Post-industrial revolution, Tsuen Wan's population increased rapidly. At the same time, the number of factories in the area diminished markedly. In 1961, Tsuen Wan's population numbered approximately 85,000 (Lai and Dwyer 1964: 153); by 2011, it was over 300,000,[5] making the district one of the largest and most densely populated and built regions in the New Territories. When I asked one of my informants to describe Tsuen Wan to me, she replied:

> When you talk about the New Territories, you think of places that are more spacious, and you need to travel a longer time to get there. But in fact, Tsuen Wan is not that kind of New Territories. I don't even think of Tsuen Wan as part of the New Territories.

Her comment clearly shows how parts of the "urbanesque" are slowly being subsumed by the expanding cityscape of Hong Kong. Moreover, the Buddhist mindfulness center, which is located inside an industrial building, is a good example of just how densely settled Tsuen Wan has become, a degree of denseness that has culminated in the need for new and already established religious centers to move or relocate inside non-religious buildings like this industrial building. As I will show in this chapter, this repositioning influences people's experiences of religious buildings, especially when such buildings, at first glance, lack overt religious characteristics. I will elaborate on these experiences later in this chapter.

Second, the anthropological insights I provide in this chapter will prove valuable to studies of how religious space is produced and negotiated, and of how a religious movement's visibility or invisibility relates to issues of self-representation and aesthetics. Other contributions on this topic have focused either on shifting ascriptions of religious spaces by religious movements creating new perceptions of the holy and the secular, or on how new religious movements locate themselves in densely populated, multicultural landscapes. I propose to explore questions of the production and negotiation of religious spaces from the viewpoints of individuals who visit these spaces, and who produce and negotiate them. Religious studies scholar Robert Orsi argues that researchers have to pay attention "to religious meanings, to multiplicities, to seeing religious spaces as always, inevitably, and profoundly intersected by things brought into them from outside, things that bear their own histories, complexities, meanings different from those offered within the religious space" (Orsi 2005: 167).

Orsi stresses the importance of studying people who come to the buildings and project their own ideas and religious meanings into religious spaces. Because the significance they ascribe to religious spaces might differ from those of official religious organizations or designers, it must be taken into consideration when researching religious spaces in urbanesque environments. For individual practitioners, places "not only *are*, they *happen*" (Casey 1996: 127, original emphasis) rendering the production and negotiation of spaces ongoing processes. It is only by looking at individuals' responses to buildings that one can fully understand the lived experiences and practices of religion (Engelke 2012).

Third, building on the meanings ascribed to religious buildings by individuals, this chapter ties into the recent debate on the materiality of religion and the question of how religion mediates "between the levels of humans and some spiritual, divine, or transcendental force" (Meyer et al. 2010: 210). Religion as a medium bridges the gap separating these levels (see also De Vries and Weber 2001). In the past, research into religion emanated from a perceived dualism between spirit and matter; this dualism was primarily based on Protestant notions of religion which placed "personal experience and immediate encounters with the divine at the core, and [regarded] form and (church) structure as secondary" (Meyer 2011: 28). Consequently, religion and media have been seen to be in opposition to each other. New insights have demonstrated the compatibility, even the integrality of both. Anthropologist Talal Asad states that "the materialities of religion are integral to its constitution" (2001: 206). Even though belief is still considered a crucial part of religion, it is not the only part. In the words of Webb Keane, "*It [religion] must still take material form* . . . Ideas are not transmitted telepathically. They must be exteriorized in some way, for example, in words, gestures, objects, or practices, in order to be transmitted from one mind to another" (2008: 230, original emphasis). Hence, religious media are intrinsic in developing relations between persons and the divine.

According to scholars active in this bourgeoning field of research, the study of religious materiality should begin "with the assumption that things, their use, their valuation, and their appeal are not something *added* to religion, but rather inextricable from it" (Meyer et al. 2010: 209; original emphasis). To research this in relation to lived Buddhism and Catholicism in Hong Kong, I will focus on two specific religious spaces. I have started my investigation with the media, instead of looking at which ones are important for my informants. This has allowed me to explicate the intriguing relationship between specific places and religious practices in the densely populated and densely built Hong Kong. The media I focus on here are not the new media technologies that Birgit Meyer examines in her recent research project *Modern Mass Media, Religion and the Imagination of Communities* (2006), but "old" forms of media that are ascribed "modern" meanings in the region of Hong Kong, meanings which extend beyond the boundaries of buildings and of the Hong Kong region. These are, in effect, meanings that reflect the presence of the global in the local. Meyer and colleagues state that religious sites become places "where people go to find themselves part of something larger" (Meyer et al. 2010: 209). In this case, these places include the global Universal Catholic Church, and the global network of Buddhism reflected by Buddhist masters coming to Hong Kong from different Asian and non-Asian countries.

Catholic and Buddhist spaces

During my fieldwork in Hong Kong, I often attended Sunday services in the Annunciation Church. The two parish priests attached to the church introduced me to members of the congregation. During my interviews with some of the church parishioners, which mostly took place in the church itself, I often asked them to reflect on Catholic churches in Hong Kong, and on the affiliation they felt with these Catholic spaces. Most of my informants emphasized the differences they perceived between Catholic and Protestant churches in Hong Kong.

One of my interviewees, a man in his mid-fifties named Albert, had been baptized as a Catholic only a few years earlier. Seated alongside me on a plastic chair in the covered courtyard of the church, he told me how he grew up in a Protestant family, and went to a Lutheran primary and a Protestant secondary School. One particular Sunday, he accompanied his girlfriend (now his wife) to a Catholic Mass, an experience that set him on the path to becoming a Catholic. Based on his experiences of Protestant and Catholic churches, Albert said that he appreciates Catholic churches as they are "all the same, because they have their own tradition and system." What Albert appreciates about Catholic churches is how they represent a relationship with God that a person can both develop and nurture in such buildings. For him, developing such a relationship is not as easily achieved in Protestant churches, which are located "in small houses, in large flat buildings, everywhere" and which seem to Albert to be too diverse to convincingly represent a universal relationship between people and God.

Sarah, a 24-year-old Opus Dei member, expressed a similar view. Sarah was baptized in the Annunciation Church in 2001. Even though she appreciates the Annunciation Church as the building she religiously grew up in, it is not the only church she attends. Depending on how busy her schedule is on Sundays, she attends either the Annunciation Church or a church close to her home in Sha Tin, an area into which she has recently moved. If she is out, she would try to attend afternoon Mass in a church in Prince Edward. Sarah said that she is not too concerned regarding which church to go to, as "my church is everywhere. Even Rome is my church." She added that for her, "this is the nice thing compared to Protestant churches, which all have their own buildings." By this, she meant that all of the Protestant buildings are different and part of separate Protestant organizations. For Sarah, it is important that the Catholic Church is one universal organization; according to her, all churches within and outside of Hong Kong are representations of this universal Church. This is not to deny that there are Protestant denominations that have links with international networks; it is more that my Catholic interlocutors do not see them as universally organized in the way that the Catholic Church is.

Both Albert and Sarah said that they value Catholic churches as representations of the Universal Catholic Church. Each church offers religious experiences of the relationship between them and God. This view is shared by Kelly, a young woman who was baptized a Catholic in 2013. Before being introduced to Catholicism by her boyfriend, Kelly used to attend a Protestant church located in a residential complex. Part of what attracts Kelly to Catholic churches is the architecture and aesthetics of the buildings, which "give me the feeling of peace, calmness and quietness." In contrast, Protestant churches "might only be a single room. You just walk in a building and only one floor will be the gathering room or the church. It feels like going to my friend's house." In Catholic churches, Kelly feels that she needs to alter her behavior to fit the environment.

The above narratives show that my informants value a Catholic church building as a point of representation of the Universal Church, and as a medium through which they can uphold and develop their personal relationships with God. It follows from this that Catholic practitioners may have similar religious experiences in each Catholic church they attend. Undoubtedly, there are differences in church buildings, and these are appreciated. For example, some of my informants commented on the fan-shaped layout of the chapel in the Annunciation Church, the mosaic at the back of the church, and the open courtyard which offers a space in which to gather after Sunday Mass. Moreover, the qualities of the individual parish priests were acknowledged and appreciated. One of the priests at the Annunciation Church, for example, is valued as being focused on the meditative aspects of prayer. However, these features do not determine the meaning ascribed to churches, nor do they play a role in influencing my informants vis-à-vis which church they would attend. Sarah opts for the church that is closest to where she is at particular times on Sundays. Albert chooses to attend mass at the Annunciation Church because, according to the Hong Kong Catholic Diocese, geographically he belongs to that parish. In conclusion, while the above narratives show that Catholic churches in general are important media in the religious practices of the Catholics of Hong Kong, this does not count for specific church buildings (including their geographical and aesthetic characteristics) such as the Annunciation Church.

Compared to my Catholic informants, Buddhist practitioners in Hong Kong tend to ascribe considerably less importance to specific religious buildings. Most of my Buddhist informants travel from one Buddhist temple or center to the next, sometimes even to non-Buddhist spaces, in search of the teachings they need for their religious experiences and practices. For them, it is not the building but the master that is the medium through which they seek to cultivate their individual religious natures.

One of those who support this approach is Teresa, a practitioner in her late thirties. I met Teresa through a program offered by the temple of which the monk of the Buddhist mindfulness center had earlier been abbot. Now that this Buddhist monk has his own center in the TML tower, Teresa often goes to this building to listen to the *dharma* teachings and to practice. Teresa had started her

Buddhist learning a few years earlier in the Chi Lin nunnery, where she attended a foundational course every Sunday for half a year. She had visited a small Buddhist temple on Lantau Island, joined a retreat on Cheung Chau Island, and was at the time I met her thinking of undertaking a Master's degree of Buddhist Studies at Hong Kong University. She had joined several meditation camps held at different places, led by international masters such as the Australian monk Ajahn Brahm and the Vietnamese monk Thich Nhat Hanh. She had even traveled to places outside of Hong Kong such as the Dharma Drum Foundation in Taiwan. For Teresa, all of the Buddhist places she attends are places where she can listen to the teachings of masters, and practice mindfulness and meditation techniques. A temple or Buddhist center for her is merely a physical place of shelter or a place in which to gather with others.

Teresa's experience was echoed by Francis, who I met during the celebration of Buddha's birthday in 2013. During that celebration, he had taken refuge in the Three Jewels (the first step in embarking on the path of Buddhism) under the tutelage of the Malaysian monk, in the temple where the latter had previously been abbot. In 2011, prior to his taking his refuge, Francis had stayed at the temple for a while, sorting out his personal problems with the help of the monk. He liked staying in the temple, not merely because of its location, but primarily because of the presence of the monk. When I asked Francis in 2014 about the new Buddhist mindfulness center in the TML Tower, he replied that the center "is just a place where we can come together." I asked Francis if he missed attending teachings in the temple where he had earlier stayed, as that was the temple where he officially became a Buddhist. He answered: "I loved that temple because of the master's warmth. Once he withdrew, it became just a building."

As is the case with the Hong Kong Catholic practitioners and their appreciation of the specific features of churches, my Buddhist informants likewise value certain Buddhist temples or centers for their architecture or aesthetics. Buddhist temples offer them certain experiences that can prove helpful in the cultivation of their Buddha-nature. For example, Teresa said that all Buddhist places give her the same feeling, "You have to be careful and everything, because you are responsible for what you are doing. So you will clean up after yourself. You have to open up to be more mindful." However, buildings are not the main media through which Buddhist practice and cultivation are generated. Instead, the main media are the Buddhist masters, who are the repositories of the teachings.

The urban and the urbanesque in Hong Kong

The above observations need to be interpreted against the background of the region of Hong Kong, mainly its increasing density. The Hong Kong region is bounded by water and the border with mainland China, making it impossible to extend geographically. Moreover, the region is well known for its increasing population pressure and limited supply of flat, buildable land (Chan et al. 2002). Even though Hong Kong has a land area over ten times that of Paris and twice the size of Singapore, it has only 17,600 hectares of buildable land (Lau et al. 2005: 529) to house a population of over 7.2 million.[6] Thus, "building density has been one of the widely accepted solutions used to satisfy the intense demand for urban activities in Hong Kong" (Chan et al. 2002: 162). Shortage of land has resulted in a high degree of Multiple Intensive Land Use (MILU): "intensification of land use through mixing residential, commercial and other uses at higher densities at selected urban locations" (Lau et al. 2005: 527). Lau and colleagues' scanning of the area of Hong Kong revealed a high number of MILU buildings in residential areas in the region:

> The overlapping, and mixing of functions in buildings may make one puzzled as to whether the building should be classified as commercial or residential. This symbiosis of privately controlled space and "borrowed" (commercial) space underpins a significant cultural trend in Hong Kong essential for vitalizing multi-use development (Lau et al. 2005: 553)

Hence, due to the increasing density of Hong Kong, buildings often become multi-purpose. This not only applies to the urban part of Hong Kong, but also to the wider urbanesque region.

As the example I provided at the beginning of this chapter makes clear, increasing density influences religious spaces in Hong Kong. Especially since the Handover of 1997, many religious organizations have no chance to building new religious centers in urbanesque Hong Kong, as space is already over-built upon or too expensive to buy. The only available option is to establish new religious centers in already existing residential, industrial, or commercial buildings. One exception is the recently built Tsz Shan Buddhist monastery in Tai Po District (New Territories). However, the purchasing of the land and the building of this new temple were only possible because it was funded by Li Ka-shing, the richest tycoon in Asia. For other religious organizations, a similar act would be simply impossible.

The situation in Hong Kong, therefore, reflects a limitation of the concept of the "urbanesque," as indicated in the introduction. The "urbanesque" is a useful term for analyzing religious practices in urban settings that have the ability to spread outward geographically, and to incorporate the semi-urban or semi-rural in their cityscapes. One example of this particular process appears in the edited volume *Public Religion and Urban Transformation*, in which different religious communities in metropolitan Chicago are described (Livezey 2000). However, what the example of Tsuen Wan makes clear is that Hong Kong's increasing density not only affects life in the central business district, traditionally seen as "the urban city" (primarily Central, Admiralty and Wan Chai), but likewise impacts on areas that are part of the "urbanesque" such as areas in the New Territories. Today, even in these areas, land is scarce and expensive. As a consequence, many religious organizations are not taking their place in "greater" or "metropolitan" Hong Kong; instead, they are being pushed inside buildings.

Concluding remarks

In this chapter, I have looked at two specific media, namely the Annunciation Church and a Buddhist mindfulness center, both of which are situated in Tsuen Wan. By looking at the meanings ascribed to these buildings, I have specified their function for religious practice. In the second half of this chapter, I have emphasized the increasing density of Hong Kong, density that must be taken into consideration when explaining the functions of the two media. For visitors to the Annunciation Church, it is not the specific church that is of primary importance in their religious practice; rather, it is the representation of the Universal Catholic Church and their personal relationships with God that are manifested in the building. With regards to my Buddhist informants, a specific place has little significance, as it is merely seen (in the words of the Buddhist master alluded to above) as a "place of shelter" where the laity can practice or listen to *dharma* teaching. Thus, the medium for practice is not a specific site: it is a master.

The narratives show that the situatedness or specific aesthetics of a religious building play an irrelevant role in religious practice. The two religious spaces are merely seen as "middle grounds"; "something *through which* something else is communicated, presented, made known" (Engelke 2012: 227–8, original emphasis). As far as my Catholic informants are concerned, it is of little relevance where a Catholic church is located or what it looks like. In effect, it matters little which church one attends, either in Hong Kong or in the rest of the world, given that in each church a person can have a similar personal religious experience. The only importance here is that it should be a Catholic church, not a Protestant church. Buddhist practitioners see buildings merely as places to practice and shelter in, places where an inherent Buddha-nature can be cultivated. Religious practice is not dependent upon the building, but on what happens inside that building; for example, one's relationship with the Buddhist master and his teachings, or the specific practices that take place in the building. Again, it is seemingly of no relevance where this religious space is located, or whether the architecture of the building looks impressive or not, an important observation in the dense urbanesque environment of Hong Kong.

These observations led to an acknowledgement of the "bundling" character of media such as religious buildings: "The contingent coexistence of an indefinite number of qualities in any object, which always exceeds the purpose of the designer" (Keane 2008: 230). Meyer suggests that "what a medium is and does is not intrinsic to the medium itself, but subject to social processes that shape religion mediation" (2011: 31). Thus, the meaning implicit in a religious medium is not intrinsic solely to that medium, but something ascribed to it by practitioners. This is especially apparent in the case of the Annunciation Church, which was, as suggested earlier, originally designed as an Italian church, but is not directly appreciated as such. In such cases, the materiality of the church is "reduced to the role of medium" (Engelke 2012: 227) as it comes to represent something that is not inherent in it.

Regarding the Buddhist mindfulness center, the story is more complex. Even though the center is a medium in which masters and laity can gather and practice together, it is not perceived as the most important medium in religious practice. Unlike the Annunciation Church, a temple is not necessarily a place that connects people to a divine being located outside of themselves. It is merely a place of shelter for body or mind. This relates to what I see as a limit to the recent debate surrounding the materiality of religion and of religious media. To date, most of the research undertaken in this field has focused on monotheistic religions and mediations; for example, the mediating function of cassette tapes for Islamic practice (Hirschkind 2006), Pentecostal mediations in Ghana and other African countries (Asamoah-Gyadu 2004; Meyer 2010), and/or Evangelical print in historical Britain (Morgan 2011). To date, studies of other religious traditions have been few; examples include the research into media in Aboriginal Australia (Ginsberg 2006), the use of photographs in India (Pinney 2004), and the functions of media in Hinduism (Engelke 2012). In all of these studies, media function to bridge the gap between people and the transcendental or divine that is located outside of the people. Hence, the studies emanate from the existence of a transcendental force that can be accessed through mediation. For example, even though in Christianity one can become a perfect image of God, one can never become God. In contrast, Buddhists regard the divine as located inside a person. In other words, each person has a Buddha-nature inside him/herself that can be cultivated. Consequently, every person has the potential to become the Buddha (literally the "enlightened one") by listening to the words of the Buddha. Unlike Christianity or Islam, the divine is not something located outside of humans, a phenomenon that influences how mediations figure in Buddhism compared to other religions. It would be interesting to explore how mediations figure in religions in which the divine is located within each person. In this chapter, I have attempted to pursue this line of inquiry.

Acknowledgments

I thank Dr. Jaap Timmer for his assistance and support during my PhD program and for his comments that greatly improved this book chapter. I am also thankful for Dr. Estelle Dryland for her meticulous editing of my manuscript. Finally, I thank the editors of this book for their invitation to write a contribution and for their comments on earlier versions of the manuscript.

Notes

1 The TML building is advertised as being an industrial building; however, it is not (as the name might suggest) a factory or manufacturing building. Industrial buildings in Hong Kong are primarily used for other related purposes. They are "mainly for logistics facilities, light industry and back office/cheap office locations, as well as [for] specialist factory operations and rapidly growing data center developments" (Benson 2015).

2 This chapter builds on the ethnographic fieldwork that I conducted in Hong Kong between 2012 and 2014. During the fifteen months I spent in Hong Kong, I investigated

the ways in which Hong Kong Buddhist and Christian practitioners relate to contemporary socio-cultural and political changes in the region.

3 More precisely, *Vihara* is the Sanskrit and Pali term for a Buddhist monastery.

4 Source: http://www.discoverhongkong.com/eng/cruise/shore-excursions/day-trips/new-territories-tour.jsp#ixzz3SFlqyqsm. Emphasis added; accessed October 1, 2014.

5 Source:http://www.census2011.gov.hk/en/district-profiles/tsuen-wan.html; accessed March 18, 2015.

6 Source: http://www.censtatd.gov.hk/hkstat/sub/so20.jsp; accessed March 18, 2015.

References

Asad, Talal (2001), Reading a Modern Classic: W. C. Smith's "The Meaning and End of Religion," *History of Religions*, 40(3), 205–22.

Asamoah-Gyadu, Kwabena (2004), Pentecostal Media Images and Religious Globalization in Sub-Saharan Africa, in Peter Horsfield, Mary Hess and Adan Medrano (eds), *Belief in Media: Cultural Perspectives on Media and Christianity*. Aldershot: Ashgate.

Benson, Darren (2015), Why Hong Kong Needs an Entirely New Approach to Industrial Property. *South China Morning Post* (January 23), http://www.scmp.com/property/hong-kong-china/article/1133788/why-hong-kong-needs-entirely-new-approach-industrial.

Casey, Edward (1996), How to Get from Space to Place in a Fairly Short Stretch of Time: Phenomenological Prolegomena, in Steven Feld and Keith Basso (eds), *Senses of Place*. Santa Fe, NM: School for American Research.

Catholic Truth Society (2013), *Hong Kong Catholic Church Directory 2014*. Hong Kong: Catholic Truth Society.

Chan, Edwin, Bo-sin Tang and Wah-Sang Wong (2002), Density Control and the Quality of Living Space: A Case Study of Private Housing Development in Hong Kong, *Habitat International*, 26: 159–75.

De Vries, Hent and Samuel Weber (2001), *Religion and Media*. Stanford, CA: Stanford University Press.

Engelke, Matthew (2012), Material Religion, in Robert Orsi (ed.), *The Cambridge Companion to Religious Studies*. Cambridge: Cambridge University Press.

Ginsberg, Faye (2006), Rethinking the "Voice of God" in Indigenous Australia: Secrecy, Exposure, and the Efficacy of Media, in Birgit Meyer and Annelies Moors (eds), *Religion, Media and the Public Sphere*. Bloomington: Indiana University Press.

Hirschkind, Charles (2006), *The Ethical Soundscape: Cassette Sermons and Islamic Counterpublics*. New York: Columbia University Press.

Keane, Webb (2008), On the Materiality of Religion, *Material Religion*, 4(2): 230–31.

Lai, David and D. J. Dwyer (1964), Tsuen Wan: A New Industrial Town in Hong Kong, *Geographical Review*, 54(2): 151–69.

Lau, Stephen, R. Giridhara and S. Ganesan (2005), Multiple and Intense Land Use: Case Studies in Hong Kong, *Habitat International*, 29: 527–46.

Livezey, Lowell (ed.) (2000), *Public Religion and Urban Transformations: Faith in the City*. New York: New York University Press.

Meyer, Birgit (2010), Aesthetics of Persuasion: Global Christianity and Pentecostalism's Sensational Forms, *Global Christianity, Global Critique, Special Issue of South Atlantic Quarterly*, 109: 741–63.

Meyer, Birgit (2011), Mediation and Immediacy: Sensational Forms, Semiotic Ideologies and the Question of the Medium, *Social Anthropology/Anthropologie Sociale*, 19(1): 23–39.

Meyer, Birgit, David Morgan, Crispin Paine and Brent Plate (2010), The Origin and Mission of Material Religion, *Religion*, 40: 207–11.
Morgan, David (2011), Mediation or Mediatisation: The History of Media in the Study of Religion, *Culture and Religion*, 12(2): 137–52.
Orsi, Robert (2005), *Between Heaven and Earth: The Religious Worlds People Make and the Scholars who Study Them*. Princeton, NJ: Princeton University Press.
Pinney, Christopher (2004), *Photos of the Gods: The Printed Image and Political Struggle in India*. London: Reaktion Books.

7 Spiritualizing the suburbs

New religious architecture in suburban London and Vancouver

Claire Dwyer

Introduction

On July 25, 2010, a spectacular new religious building, the Thrangu Tibetan Monastery, celebrated its grand opening in Richmond, a suburban municipality south of Vancouver in British Columbia

Heralded by its founders, as "a miniature Tibet" and the "first traditional-style" Tibetan monastery in Canada, this elaborate, purpose-built place of worship joined a diverse range of other religious buildings along the Number 5 Road on the edge of the city, part of a multicultural suburban landscape known popularly as "Highway to Heaven" (see Dwyer et al. forthcoming). In this chapter, I discuss the construction of the Thrangu Tibetan Monastery alongside two other recent, purpose-built, religious buildings constructed by diasporic migrant populations in the suburbs of London: the Jain Deraser in Potters Bar, and the Salaam Centre, currently under construction in Harrow. I argue that, in common with a range of recently constructed religious buildings in Europe and North America, these buildings are evidence of the ways in which networks and trajectories of transnational migration are shaping new geographies of faith in suburbia (Dwyer et al. 2013a). Such buildings offer a novel means to analyze how the contemporary city is "spiritualized" involving a creative materialization of religious space by transnational migrant religious communities, which, unlike many new migrant communities, are located in the suburban fringe rather than in the center of the city.

This chapter begins with a brief discussion of the challenges for new migrant faith communities in finding space for worship in European and North American cities. This is an account of improvisation and adaptation of existing spaces as well as a negotiation of urban planning restrictions and regulations. The construction of new purpose-built places of workshop may be the ultimate goal of a migrant faith community, but it requires time and investment as well as the ability and social capital to negotiate bureaucratic municipal barriers. If academic attention has focused primarily on such procedural processes, this chapter reflects particularly on the architectural forms of new religious buildings and the extent to which such buildings offer new forms of sacred geographies in the city, and in particular a distinctive engagement with suburban landscapes. Such religious

Figure 7.1 Thrangu Tibetan Monastery. Photo by Claire Dwyer.

buildings are shaped by transnational trajectories and offer a distinctive, hybrid aesthetics. Widening the scope of this volume from cities to suburbs, I also reflect on the specificity of the suburban as the site for some of the most significant new religious buildings in the UK and North America. Three different examples from my research are drawn on to explore the making of new religious spaces in the suburbs by different migrant groups. First, is the Thrangu Tibetan Monastery in Richmond, located in the suburban fringe within a designated planning zone for places of worship. The next two examples are from London's suburbs, including a Jain Temple, located in the outer London commuter suburbs within the "greenbelt," and a new Muslim center currently under construction in the northwest London suburb of Harrow.

Migration and religion: building places of worship

Studies of the significance of religion for migrants to both Europe and North America have often questioned the possibilities and barriers for migrant groups in their participation in ethno-religious congregation and practice and questioned whether religion is an integrative process for immigrants (Foner and Alba 2008). However, increasingly scholars recognize the transnational practices of migrants. This includes the ways in which religious transnationalism involves the mobilization of new networks and flows of finance, social remittances, objects and people across national borders (Levitt 2007) as transnational religious spaces and communities are formed in different contexts (Ebaugh and Chaftez 2000; Sheringham 2013). While there has been a recent resurgence of interest in religion in mainstream European urban studies (Becker et al. 2013) the architectural and material cultures of migrant faith communities has been less well documented than the sociological aspects of religious transnationalism. While it is true that space for worship for migrant religious groups is often ad hoc and provisional, the establishment of more permanent buildings or spaces for religious practice is an important barometer of multicultural planning and the diversity in urban space (Fincher et al. 2014).

In a typology of "new religious landscapes" in the UK, Peach and Gale (2003) trace a typical cycle of development of places of worship for minority faith groups. They suggest that most begin in domestic spaces, usually without formal planning permission. Expansion requires the purchase and conversion of existing buildings, including the frequent conversion of churches to temples and mosques, or the use of industrial premises. Minority faith communities face not only the challenge of purchasing buildings, but also getting planning permission for their conversion or reuse. Research suggests that minority faiths, particularly Muslims, face particular difficulty in gaining planning permission for new worship spaces (Dunn 2005; Ehrkamp 2012; McLoughlin 2005; Gale 2004, 2008; Isin and Siemiatycki 2002). Opposition to new religious buildings is usually expressed indirectly, focusing on issues of noise or disturbance or concerns about increased traffic or car parking (Gale and Naylor 2002). Gale (2005), in a discussion of mosque building in Birmingham, emphasizes the need for worshippers to have strong local, political networks and good knowledge of the institutional context. Opposition may also sometimes be raised against forms of architecture, which are seen as intrusive, "alien," or unfamiliar (Naylor and Ryan 2003), although new religious buildings may also be celebrated as sites of multiculturalism, as Zavos's (2009) discussion of the Hindu community's promotion of the Swaminaryan Temple in Neasden suggests.

As the sites discussed in this chapter reflect, such locational conflicts may increasingly take place in the suburbs. This reflects not only the suburbanization of longer established immigrant groups, but also the increasing role of suburbs as "gateways" for new immigration (Ehrkamp and Nagel 2012). It is also a function of the distinctive geographies of suburban space (Dwyer et al. 2013a). In some cases, such as the spectacular purpose-built Shri Swaminaryan Temple in north-west London, its location in the inner suburbs of Neasden followed difficulties in gaining planning permission in more prominent and affluent locations. Similarly, the Mohammedi Park Mosque is located on a canal-side industrial estate in nearby Northolt (Eade 2011). Elsewhere, like the model of the suburban mega-church (Wilford 2012), the suburban fringe provides more expansive opportunities for purpose-built places of worship such as the BAPS Shri Swaminaryan temples in Toronto and Chino Hills, California (Kim 2010), or the Ahmadiyya Mosque in Vaughan, north of Toronto (D'Addario et al. 2008). The transitional geographies of suburban areas may also provide provisional and improvised spaces of worship like the former warehouses and industrial buildings used as Hindu temples in northern Toronto (Hackworth and Stein 2012), or African Pentecostal churches in parts of east London (Krause 2011).

Discussion of the architectural styles of new religious buildings has often been limited to the extent to which such buildings are consistent, or unfamiliar, in the surrounding cityscape. Thus opposition to the Mohammedi Park Mosque highlighted its "alien" architecture (Eade 2011), while Naylor and Ryan (2002) recount the history of one of the earliest mosques in Britain, the London Fazl Mosque built in 1926 in Southfields in the southwest London suburb of Putney and variously described as "exotic" and "Orientalist picturesque." Peach and Gale (2003: 286) argue that the aesthetics of Islamic, Sikh and Hindu architectural styles challenge the "normalizing language of planning discourse" in their "architectural forms, building materials and decorative colours." In contrast, Saleem (forthcoming) traces a history of the "British Mosque," linking trends in early mosque building to other architectural histories. Thus the influence of Lutyens Modernism is recognized in the Southfields mosque, while the Mughal-inspired domes of the first British mosque in the London suburb of Woking in 1889 echo other Orientalist buildings such as Brighton Pavilion (1825) and Leighton House (built in Holland Park in 1864). Salaam raises the question of what form newly built British mosques should take, lamenting the "indiscriminate" lifting of architectural styles and Islamic motifs to a new context.

This chapter reflects on the creation of new purpose-built places of worship by migrant religious groups. Offering a comparison of three new purpose-built religious spaces, constructed in suburban locations in London and Vancouver, it suggests that all of these buildings offer evidence of re-invention and re-creation. Even those buildings, which are narrated as faithful renditions of "traditional" or "native" architectural forms, are translated through migration to produce new hybrid forms of religious worship space. In contrast to prevailing accounts of new immigrant religious buildings in the suburbs, which emphasize the incongruity and disruption of such new architectural forms, this chapter suggests that there is scope for celebrating the imaginative and hybrid religious architectures which have begun to emerge. The first case study, located in Richmond, Vancouver, traces the emergence of a building, which is understood as a careful copy of a traditional Tibetan temple but whose construction belies a more complex assemblage.

Thrangu Tibetan Monastery, Richmond, Vancouver[1]

The opening of the Thrangu Tibetan Monastery in Richmond, Vancouver on July 25, 2010 was attended by local media, politicians, and worshippers who included Hong Kong Chinese migrants, Tibetan émigrés and Western converts to Tibetan Buddhism.

The Thrangu Monastery was the culmination of a long-term vision of the spiritual leader of the Thrangu Vajra Vidha Buddhist Association, Khenchen Thrangu Rimpoche, to build a Tibetan monastery in North America. Adherents to this branch of Tibetan Buddhism had grown in North America since the 1970s when

Thrangu Rimpoche was based in Nova Scotia. However, the project was realized through the specific trans-Pacific flows, which have shaped the city of Vancouver since the 1980s with the settlement of Hong Kong Chinese and other transnational Asian migrants (Ley 2010). Almost 60 percent of the population of the suburban municipality of Richmond were recorded as "immigrants" in the 2006 Census, many from Hong Kong, Taiwan and mainland China. The Thrangu Vajra Vidhya Association has built up an important following in Taiwan, Hong Kong, and amongst the Chinese community in Vancouver; and the monastery was made possible by the munificence of one family of donors, owners of the Hong Kong-based Henderson Holdings property empire.

The Thrangu Monastery stands on Number 5 Road, on the eastern boundary of the multicultural suburb of Richmond and adjacent to the major Highway 99, where there are currently more than twenty religious buildings. These include two mosques, eight churches (many Chinese speaking), four religious schools, two other Buddhist temples, two Hindu temples, and a Sikh gurdwara. This diverse religious landscape is the outcome of a unique piece of suburban planning legislation, which united the preservation of agricultural land (the British Columbia Agricultural Land Reserve), the prevention of urban sprawl, and the provision of land for "Assembly Use." Permission to build new religious buildings is given under strict conditions requiring the maintenance of part of the land in active cultivation. The city planners who initiated this planning policy in 1990 did not anticipate the diverse religious agglomeration, which has followed as the number of new buildings has grown, resulting in a landscape, which is celebrated by local politicians and marketed as a site for "multicultural" tourism (Dwyer et al. forthcoming).[2]

Although the land on Number 5 Road was purchased in 2004, gaining planning permission for the Thrangu Monastery was protracted. Not only did plans for farming have to be approved, but the monastery had to negotiate planning permission for a spectacular building, which sought to replicate Tibetan monasteries in Nepal. Eventually permission was secured to build to a height greater than eleven meters higher than that usually permitted in this "Assembly District"; it was allowed in recognition that the requested roof-top cupola was "an important component of the vernacular architectural vocabulary of Tibetan temples."[3] This adherence to an "authentic" Tibetan style of building was central to the narrative constructed around the new temple, which was defined as the first "traditional" Tibetan temple in North America. Yet research with the architects and construction company, as well as the monks and local Buddhist members involved in the building, reveals the complex process by which the new building was realized.

Matthew Cheung, the architect chosen for the Thrangu Monastery, had previously built Buddhist temples in Vancouver (including the Japanese International Buddhist Temple) but had little knowledge of Tibetan Buddhist architecture. He recommended Kindred, a local construction company, which specialized in custom built projects such as luxury housing in Western Vancouver, and had previously constructed another Taiwanese Buddhist center on Number 5 Road, although with a distinctively "West Coast" architecture. The architects were tasked with the "replication" or "full emulation" of the Tibetan Monastery in Nepal, yet had to find ways to achieve this within the specific local geography and within a compressed time frame. The limitations of the local geography and fire regulations required a creative response from the architects and the construction company. First, they had to find a way to reduce the weight of the building given the status of the floodplain peat sediment upon which it was to be built. Thus the building has a steel frame interior with an outer cladding of brick and stone to give an impression of solidity, as the architect explains: "you think it's all solid concrete and massive, it's not." Faced with the task of recreating the "mountain architecture" of Nepal, on a flood plain in Canada, the architects sought creative ways to add elevation to their building, with undercover parking at the rear, and entrance to the first-floor shrine hall up a flight of stairs. As the architect explained: "in a sense it's elevated . . . like the whole experience of entering the temple, you don't directly go inside. You have to climb the stairs and then go inside. We tried our best to raise it as high as possible."[4]

A second challenge involved the decorative moldings on the temple, which would traditionally be molded in situ from concrete. On the building's exterior, the window cornices were made of glass-reinforced concrete over steel members, to reduce weight. Within the interior of the monastery, only the wall behind the giant Buddha was made in the traditional way with molded concrete in situ. For other decorative moldings, Kindred employed a Hollywood set sculptor Sergi Tzchenko (Dreamcast); the decorative moldings in the shrine hall were made of fiberglass. These were then mass produced and painted "to make it look traditional." To complete the project in twenty months, rather than the "twenty years" usually required, scores of volunteers were drafted in to paint the decorative pieces.

The focus of the shrine hall is the 4-meter-tall gold-plated Shakyamuni Buddha (filled with precious offerings—scriptures, scrolls and sacred stones from more than a hundred countries) and flanked by 35 smaller buddhas of confession. Within glass cases on the side walls are a thousand small "medicine buddhas" (a devotional object of healing), while there are two hundred statues of Amitabha adjacent to the main entrance. These were the last elements of the temple to be installed and—as at earlier stages of the construction—appropriate ceremonies accompanied their installation. These ceremonies were crucial in the realization of the monastery as a sacred space.

The Thrangu Monastery is perhaps the most elaborate and spectacular of the newly built places of worship on Number 5 Road, and its completion has appeared to encourage some of the other temples along the road to apply for expansion. What is particularly interesting about the Thrangu Monastery is how its construction, although represented as an "emulation" or replication of a "traditional" Tibetan monastery, is instead a more transnational construction, realized through flows of people and information and the creative adaption of ideas to the local topography and geography. This is an innovative hybrid construction process. This hybridity is mirrored in the emerging religious life of the new monastery. Like many other transnational religious spaces, the monastery emerges as a nexus of different kinds of religious practice with worshippers from different traditions and linguistic backgrounds. For the Tibetan émigré, the temple is an important symbolic space—important for New Year celebrations and as a form of home-making—although some have demonstrated political resistance to its construction and funding. Visitors to the temple are predominantly of Chinese backgrounds, transmigrants to Canada from Hong Kong or the People's Republic of China. For Western Buddhist visitors, the dominance of Chinese languages in the organization of activities in the monastery is unfamiliar and sometimes unsettling, reflecting traditions of Buddhist worship, which are more communal and less text based.

The Thrangu Monastery's location on Number 5 Road, on the suburban fringe of Richmond, Vancouver is part of a wider transformation of this suburban landscape as planning regulations, facilitating the location of religious groups, have resulted in a diverse, multicultural religious landscape (Dwyer et al. forthcoming). This landscape is also a reflection of Richmond's emergence as a distinctive "ethnoburb," or ethnic suburb (Li 2009), shaped by recent Asian-Pacific transmigration. The Thrangu Monastery, like other Buddhist temples on Number 5 Road, is part of a wider process of transformation in Richmond, alongside suburban Chinese retail spaces like shopping malls and markets (Pottie-Sherman and Hiebert 2015). This process of ethnic change is not necessarily welcomed by all inhabitants, although the religious "exoticism" of Number 5 Road has been celebrated by municipal government and even promoted through the local tourist agency as a site of "multicultural" tourism.

Thrangu Monastery, and its surrounding religious buildings, can then be read as a form of "spiritualization" of the suburbs whereby this "edge-city" location of agricultural land has been transformed through the construction of a landscape of religious buildings, which serve a range of different religious communities. This is a materialization of the religious and the sacred in the assumed secular spaces of suburbia (Dwyer et al. 2013a). Worshippers are both locals, from this increasingly Asian-Canadian suburb, as well as faith commuters from across metropolitan Vancouver attracted by new purpose-built and expansive places of worship. These hybrid, diasporic buildings are also somewhat transformative for local populations, who largely celebrate this new landscape, although they do this primarily through a register of positive Canadian multiculturalism rather than religious or spiritual discourse. Nonetheless the "temple tours" run by the local museum are a popular way to visit these new places of worship (see Dwyer et al. forthcoming).

The Shikharbandhi Jain Deraser, Potters Bar, London[5]

The second example I want to draw on, from the suburban fringe of London, presents a very comparable narrative of hybrid religious architectural style. Shikharbandhi Jain Deraser is a purpose-built Jain temple, which opened in 2005 located in the landscaped gardens of Hook House, a nineteenth-century listed building.

The temple is the culmination of a twenty-year project for the Oshwal Association of the UK, a transnational Jain community whose members began migrating from East Africa to the UK in the early 1960s, but whose origins lie in Jamnagar, western Gujarat in India. The original building and its surrounding gardens and fields were first purchased in 1979. There then followed a period of fundraising and planning permission applications to construct the new purpose-built temple. Planning permission was eventually granted, despite the building's location in the "greenbelt" (a planning zone of restricted urban development around London implemented in 1947) on the understanding that the existing building would be retained and the new buildings sufficiently concealed within the existing landscape. The temple is described by the Oshwal Association "as Europe's first traditional *Shikharbandhi* Jain Deraser to be built on virgin land." This narrative is central to the role of the temple as an emerging site of diasporic identity for Jains in Britain and beyond (Shah 2014). The ambition was to build a "*tirth*," a site which could be approved as a sacred pilgrimage site for Jains. Traditionally a "*tirth*" would be a temple constructed in a remote region, which would acquire spiritual status and authenticity as a sacred site. Like the Thrangu Monastery, the Jain Deraser is not a simple recreation of a South Asian religious architectural form, but is instead a more hybrid architecture and landscape, which responds to the local situation and planning regime.

The temple is made of Indian pink sandstone and marble. The elaborate hand-carved pieces were carved in India and then shipped to the UK to be assembled. According to the Jain religious texts, or *Shilpashastras*, no steel or other ferrous metal can be used in the temple construction. The British architects, Ansell and Bailey, substituted a bronze compound to reinforce the laid segmental dome and columns inside the temple. The elevator for disabled access was constructed to be just one centimeter apart from the temple itself thus facilitating religious stipulations. However, there were other compromises in the underfloor heating and concealed lighting. Outside, the architects had to contend with northern European weather conditions, coating the red sandstone and marble walls to protect from rain damage.

The biggest challenge, however, was, like the Thrangu Monastery, in the elevation of the building. Planning for the new building was given on the stipulation that the highest post of the temple, the stone pyramid structure at its center, could not be higher than the highest point of the existing building on the site. In order to maintain the approved religious symmetry of the building and retain the height of the dome, the architects decided to excavate the site, creating a sunken landscaped garden in which to locate the temple. This formal ornamental garden, which is not typical for Jain temples in South Asia, suggests a hybrid design uniting the traditions of a formal European landscaped garden with Jain statuary laid out in a geometric formation of the *Triloka* or Jain sign for the cosmos. Planting in the landscaped gardens also make direct reference to the diasporic history of the Jain community. Beyond the ornamental garden is a new grove of 52 eucalyptus trees planted to recognize the 52 villages in Halar District, western Gujarat, in India from which Oshwal Jains trace their ancestry. Although eucalyptus is native to Australia rather than South Asia, the species has recently been promoted in Kenya to aid reforestation. The garden thus emerges as a means to create a form of European Jainism, important for the second-generation diaspora (Shah 2014).

Following its opening in 2005, the Jain Temple in Potters Bar has quickly become an important sacred pilgrimage and worship site for Jains across the UK, and internationally. The temple and the assembly halls, located in the main house, are used frequently for weddings and religious festivals. The temple has also worked hard to ensure the space is open to visitors and to educate a wider public about their faith. The temple is a regular site for school and church groups and more casual visitors are always welcomed, if infrequently. The initial proposal to build the temple prompted fierce opposition from local residents who were opposed to a new, non-Christian, religious building in their semi-rural landscape and were concerned about noise and traffic. Like the buildings in Richmond, Vancouver, the Jain Temple's location, near the London's M25 arterial motorway, is visited by car-driving "faith commuters" who travel some distance to the temple. Its "edge-of-city" location in a semi-pastoral landscape means that this temple is largely hidden from view and often unknown to locals and the wider public. Nonetheless, like the nearby Bhaktivedanta Manor, the main center for the ISKCON (Hare Krishna movement) in Letchmore Heath (Nye 2000), the Jain Temple is slowly becoming assimilated into the increasingly diverse suburban environs of London's landscape.

The Jain Temple in Potter's Bar offers a parallel example to the Thrangu Monastery; both attempt to replicate religious architecture from elsewhere in a new suburban landscape. While both have had to marry domestic planning regulations with religious stipulations and practice, these are somewhat different suburban settings. Unlike the Thrangu Monastery, the Jain Temple is more isolated, in a semi-rural area on the edge of the commuter belt; it has been able to adapt this English pastoral setting in the transnational architectures and landscaping of the new temple. What has been created is a new and distinctive South Asian spiritual landscape in suburban London.

The Salaam Centre, Harrow, London[6]

The final example in this chapter is of a new building, under construction in the northwest London suburb of Harrow. While the two previous examples are of new religious spaces, which are hybrid architectural and landscape forms, this is a building, which adopts a more self-conscious orientation towards a new and distinctive form of British religious architecture. The Salaam Centre is a new, purpose-built religious building commissioned by an Ithan'ashari community of Shia Muslims of East African ancestry. The Ithan'ashari Shia Muslims in Harrow are a Khoja community who trace their ancestral linkages to Gujarat in India, but like the Oshwal Jains, are "twice migrants" (Bhachu 1985), who came to Britain in the 1960s and 1970s. They came often as students and their permanent settlement was prompted by the expulsion of Asians from Uganda in 1974. Like the Jains, many settled in the suburbs of northwest London and over time established sites of community and worship. Premises in North Harrow were first purchased from the local council in 1990 and worship and community activities took place in an old pre-fabricated building built as a restaurant for military personnel. Planning permission for the new building was granted by Harrow Council in 2011, despite some opposition. Harrow is a diverse suburb with more than a third of the population recorded as "Asian" or "Asian British" in the 2011 census, but some opponents questioned the need for another mosque in the locality.

For the Ithan'asharis, the ambition has been to create a building, which, while serving the needs of their own Shia Muslim community, was also a wider civic and spiritual space. Choosing the name "Salaam Centre" was a self-conscious gesture intended to signify a space that was inter-faith and shared. The community also sought to construct a building that was architecturally distinctive and ambitious as an asset to the local community of Harrow; this narrative, similar to the Thrangu Monastery in Vancouver, was significant in garnering institutional and political support for the project.

The choice of the London-Barcelona architects' practice, Mangera Yvars, was therefore important in the realization of this ambition. Mangera Yvars has been at the forefront of new initiatives to create purpose-built Islamic architecture in the UK. They have designed buildings in Qatar and Kuwait, including the new Faculty of Islamic Studies for Qatar University in Doha; in the UK, they were commissioned to design the never realized Abbey Mills mosque in East London (DeHanas and Pieri 2011; Mangera 2011). Ali Mangera, of Mangera Yvars, in common with other new British-Muslim architects such as Shahed Saleem (Saleem forthcoming) has sought a new indigenous form of Islamic architecture, in contrast to what he defines as the "pastiche" style of much new mosque building in the UK—"that cartoon look, all plastic domes and minarets" (Mangera, cited in Glancey 2006).

Figure 7.2 Designs for the Salaam Centre. Photo by Mangera Yvars Architects.

Mangera's design for the Salaam Centre responded to the message of a building signifying openness and inclusion with an emphasis on light and space.

A key feature of the building is "fractal geometric patterning" incorporated into the facades, which allows daylight to filter in creating "distinctive geometric shadows" (MYAA 2010). Use of light in the building to open up the space is complemented by linked public spaces including an "agora piazza" at the entrance, designed like an "Islamic courtyard" with open-air seating and two gardens with Islamic water features. As the architect explained, the architectural form was intended to literally "open up the building, inviting visitors inside." The building's design is intended to recall the history of the community. Thus the curves of the building were intended to evoke the nomadic tented mosques of early Islam in Arabia. The migratory journeys of the Shia Ithan'ashari diaspora are incorporated more explicitly into the geometric patterning of the building's facade. Mangera Yvar's design includes sets of patterns drawn from creative designs from Persia, India and Tanzania, making a direct link to the ancestral origins of the community. However, the design's fourth patterns draw inspiration from Harrow's suburban vernacular with a reference to the motifs of English arts and crafts designer William Morris. Thus the design connects creative patterns from a diasporic journey, which connects the 1930s suburb of Harrow with other places. These journeys are echoed in the planned gardens linking the different buildings; these include both an "Islamic" garden and a "Metroland" suburban British garden, a reference to the railway networks, which produced London's suburbs, commemorated in poet laureate John Betjeman's 1973 BBC documentary, *Metroland* (Dwyer 2015).

In contrast to the previous two examples, the Salaam Centre does not seek to emulate a religious building from the homeland, rendering the sacred through a careful copy of existing and approved religious architecture. Instead, this is a making of new religious space in the city, which seeks to create a new design—but one, which explicitly engages with migratory histories and colonial and post-colonial connections.

Construction began at the site of the Salaam Centre in spring 2015, so it is too early to evaluate the role of this building within the surrounding landscape and wider community. However, it seems likely that the new building will, like the Thrangu Monastery, become a new celebrated civic landmark given its distinctive architecturally innovative style. Given the demographic diversity of Harrow, the Salaam Centre will serve a local Islamic society although it also has the potential to attract worshippers from further afield once completed.

Unlike the Jain Temple in the commuter belt of Potters Bar, or even the Thrangu Monastery, which is located right at the edge of a Canadian suburb, the Salaam Centre is located in a densely residential suburb. This is an ethnically and religiously diverse neighborhood with many other religious buildings. However, the Salaam Centre's distinctive new religious architecture provides an innovative materialization of the spiritual in London's interwar suburbs, challenging straightforward narratives of suburban secularity.

Conclusion: spiritualizing the suburbs?

This chapter has taken a distinctive perspective on this volume's wider project on spiritualizing the city. The emphasis in this chapter has not been primarily on the sociological or anthropological aspects of spirituality or religion, but on the materialization of the religious and the sacred in the built environment. Through three case studies, I have drawn attention to the increasing importance of the suburbs in changing multicultural and religious geographies of the city. The three buildings discussed in this chapter are the result of both social and demographic changes; these are associated with new transnational migration flows direct to the suburbs, and the suburbanization of older immigrant populations, coupled with the accumulation of sufficient capital by migrant generations to realize new purpose-built places of worship. In responding to the wider focus of this volume about the spiritualizing of the city, my approach has been somewhat conventional, focusing on the built form of congregational spaces of worship and highlighting the architectural significance of purpose-built, and often spectacular, new additions to the suburban landscape. However, in highlighting the suburban as an important site for the re-invention and re-creation of new religious architecture, I seek to encourage the recognition of the spiritual and the sacred in wider re-evaluations of the emerging significance of suburbia, particularly focusing on the material and the architectural (Keil 2013).

What the buildings highlighted in this chapter, and in my wider research, suggest is that the suburbs can be refigured as significant sites of transnational connectivity and post-colonial connection. Such an analysis offers a reworking of dominant readings of the suburban as an opposite to the dynamic multiculturalism of the inner city. It is an analysis, which acknowledges the increasing demographic and religious diversity of suburban areas, which are often represented as homogenous. As this chapter suggests, the location of new places of religious worship in the suburbs is usually a result of the intersection of a range of factors, including the increasing multiculturalism of the suburbs and the barriers, both political and financial, to realizing expansive new religious buildings in more central areas of the city. None of the buildings discussed in this chapter were achieved easily. They were only realized through long struggles often against local opposition. However, their eventual realization offers possibilities for transformation in the ways in which suburban communities might engage with the assumed secular spaces of the suburbs, as well as suggesting that the diversification of the suburbs is likely to increase.

Notes

1 This case study draws on research conducted by the author and colleagues Justin Tse and David Ley between 2010 and 2012, funded by *Metropolis Canada* (see Dwyer et al. 2013b; Dwyer et al. forthcoming)

2 It is striking that the municipal authorities primarily define this landscape through the secular municipal lens of Canadian multiculturalism. The promotion of Number 5 Road by Richmond Tourism as a tourist destination, which has not had much success, defines the faith communities as "ethnic" rather than religious (see Dwyer et al. forthcoming).

3 Source: Development Permit Panel Report, City of Richmond Planning and Development Department, July 16, 2007.
4 Interview with the authors, April 28, 2010.
5 Research on the Jain Temple was conducted between 2008 and 2010 with colleagues Bindi Shah and David Gilbert. See Shah et al. (2012) for a more detailed discussion of this case study.
6 For more detailed discussion of this case study see Dwyer (2015).

References

Becker, Jochen, Kathrin Klingan, Stephan Lanz and Kathrin Wildner (2013). *Global Prayers: Contemporary Manifestations of the Religious in the City*. Berlin: Lars Müller Publications.

Bhachu, Parminder (1985). *Twice Migrants. East African Sikh Settlers in Britain*. London and New York: Tavistock.

D'Addario, Silvia, Jeremy Kowalski, Maryses Lemoine and Valerie Preston (2008). *Finding Home: Exploring Muslim Settlement in the Toronto CMA*, CERIS Working Paper, No. 68. http://ceris.metropolis.net/Virtual%20Library/WKPP%20List/WKPP2008/CWP68.pdf.

DeHanas, Daniel N. and Zacharias Z. Pieri (2011). Olympic Proportions: The Expanding Scalar Politics of the London "Olympics Mega-mosque" Controversy, *Sociology*, 45(5): 798–814.

Dunn, Kevin M. (2005). Repetitive and Troubling Discourses of Nationalism in the Local Politics of Mosque Development in Sydney, Australia, *Environment and Planning D: Society and Space*, 23(1): 29–50.

Dwyer, Claire (2015). Reinventing Muslim Space in Suburbia: the Salaam Centre in Harrow, North London, in Brunn, Stanley D. (ed.), *The Changing World Religion Map: Sacred Places, Identities, Practices and Politics*. New York: Springer.

Dwyer, Claire, David Gilbert and Bindi Shah (2013a). Faith and Suburbia: Secularisation, Modernity and the Changing Geographies of Religion in London's Suburbs, *Transactions of the Institute of British Geographers*, 38(3): 403–19.

Dwyer, Claire, Justin K. H. Tse and David Ley (2013b). *Immigrant Integration and Religious Transnationalism: The Case of the "Highway to Heaven" in Richmond, BC*. Working Paper 13–06 Metropolis British Columbia.

Dwyer, Claire, Justin K. H. Tse and David Ley (forthcoming). "Highway to Heaven": The Creation of a Multicultural, Religious Landscape in Suburban Richmond, British Columbia, *Social and Cultural Geography*.

Eade, John (2011). From Race to Religion: Multiculturalism and Contested Urban Space, in Justin Beaumount and Christopher R. Baker (eds), *Postsecular Cities. Space, Theory and Practice*. London, New York: Continuum.

Ebaugh, Helen Rose Fuchs and Janet Saltzman Chaftez (eds) (2000). *Religion and the New Immigrant Congregations*. Walnut Creek, CA: Alta Mira Press.

Ehrkamp, Patricia (2012). Migrants, Mosques and Minarets: Reworking the Boundaries of Liberal Democratic Citizenship in Switzerland and Germany, in Marc Silberman, Karen E. Till and Janet Ward (eds), *Walls, Borders, Boundaries: Spacial and Cultural Practices in Europe*. New York: Berghahn.

Ehrkamp, Patricia and Caroline Nagel (2012). Immigration, Places of Worship and the Politics of Citizenship in the US South, *Transactions of the Institute of British Geographers*, 37: 624–38.

Fincher, Ruth, Kurt Iveson, Helga Leitner and Valerie Preston (2014). Planning in the Multicultural City: Celebrating Diversity or Reinforcing Difference? *Progress in Planning*, 92: 1–55.

Foner, Nancy and Richard Alba (2008). Immigrant Religion in the US and Western Europe: Bridge or Barrier to Inclusion? *International Migration Review*, 42(2): 360–92.

Gale, Richard (2004). The Multicultural City and the Politics of Religious Architecture: Urban Planning, Mosques and Meaning Making in Birmingham, UK, *Built Environment*, 30(1): 18–32.

Gale, Richard (2005). Representing the City: Mosques and the Planning Process in Birmingham, *Journal of Ethnic and Migration Studies*, 31(6): 1161–79.

Gale, Richard (2008). Locating Religion in Urban Planning: beyond "Race" and "Ethnicity," *Planning, Practice and Research*, 23(1): 19–39.

Gale, Richard and Simon Naylor (2002). Religion, Planning and the City: The Spatial Politics of Ethnic Minority Expression in British Cities and Towns, *Ethnicities*, 2(3): 387–409.

Glancey, Jonathan (2006). Dome Sweet Dome, *The Guardian* (October 30). http://www.guardian.co.uk/artanddesign/2006/oct/30/architecture.religion; accessed June 12, 2015.

Hackworth, Jason and Kirsten Stein, (2012). The Collision of Faith and Economic Development in Toronto's Inner Suburban Industrial Districts, *Urban Affairs Review*, 20(10): 1–27.

Isin, Engin F. and Myer S. Siemiatycki (2002). Making Space for Mosques: Struggles for Urban Citizenship in Diasporic Toronto, in Sherene H. Razack (ed.), *Race, Space, and Law. Unmapping a White Settler Society*. Toronto: Between the Lines Press.

Keil, Roger (ed.) (2013). *Suburban Constellations. Governance, Land and Infrastructure in the 21st Century*. Berlin: Jovis Verlag.

Kim, Hanna (2010). Public Engagement and Personal Desires: BAPS Swaminarayn Temples and Their Contribution to the Discourses of Religion, *International Journal of Hindu Studies*, 13(3): 357–90.

Krause, Kristine (2009). Spiritual Spaces in Post-industrial Places: Transnational Churches in North East London, in Michael Peter Smith and John Eade (eds), *Transnational Ties* (Comparative Urban and Community Research, Volume 9). New Brunswick, NJ: Transaction Publishers.

Levitt, P. (2007). *God Needs No Passport: Immigrants and the Changing American Religious Landscape*. New York: The New Press.

Ley, David (2010). *Millionaire Migrants: Trans-Pacific Life Lines*. Oxford: Wiley-Blackwell.

Li, Wei (2009). *Ethnoburb: The New Ethnic Community in Urban America*. Honolulu: University of Hawai'i Press.

McLoughlin, Seán (2005). Mosques and the Public Sphere: Conflict and Cooperation in Bradford, *Journal of Ethnic and Migration Studies*, 31(6): 1045–66.

Mangera, Ali (2011). The Mega Mosque, in Justin Jaeckle and Füsun Türetken (eds), *Faith in the City: The Mosque in the Contemporary Urban West*. London: The Architecture Foundation.

MYAA (2010). Planning Submission: North Harrow Community Centre: http://www.thesalaamcentre.com/application; accessed August 24, 2016.

Naylor, Simon and James R. Ryan, (2002). The Mosque in the Suburbs: Negotiating Religion and Ethnicity in South London, *Social and Cultural Geography*, 3(1): 39–59.

Nye, Malory (2000). *Multiculturalism and Minority Religions in Britain: Krishna Consciousness, Religious Freedom and the Politics of Location*. Richmond: Curzon Press.

Peach, Ceri and Richard Gale (2003). Muslims, Hindus and Sikhs in the New Religious Landscape of England, *The Geographical Review*, 93(4): 469–90.

Pottie-Sherman, Yolande and Daniel Hiebert (2015). Authenticity with a Bang: Exploring Suburban Culture and Migration through the New Phenomenon of the Richmond Night Market, *Urban Studies*, 52(3): 538–54.

Saleem, Shahed (forthcoming). *The British Mosque*. London: English Heritage

Shah, Bindi (2014). Religion in the Everyday Lives of Second-generation Jains in Britain and the USA, *The Sociological Review*, 62: 512–29.

Shah, Bindi, Claire Dwyer and David Gilbert (2012). Landscapes of Diasporic Belonging in the Edge-city: The Jain Temple at Potters Bar, Outer London, *South Asian Diaspora*, 4(1): 77–94.

Sheringham, Olivia (2013). *Transnational Religious Spaces.* Basingstoke: Palgrave Macmillan.

Wilford, Justin G. (2012). *Sacred Subdivisions: The Postsuburban Transformation of American Evangelicalism.* New York: New York University Press.

Zavos, John (2009). Negotiating Multiculturalism: Religion and the Organisation of Hindu Identity in Contemporary Britain, *Journal of Ethnic and Migration Studies*, 35(6): 881–900.

8 Multiculturalism, veiling fashion and mosques

A religious topography of Islam in Berlin

Synnøve Bendixsen

Introduction

The religious landscape of European cities has been strongly impacted by migration to Europe (Oosterbaan 2014). Material manifestations of Islam in the urban landscape—be it an increase in the number of youth wearing headscarves or purpose-built mosques constructed both in urbanesque and central urban locations—have created fierce public debates and opposition. How religious presences are regarded will have different implications for the meaning of religious performances and practices for participants in these spaces. Orsi (1999, 46) stresses that "[W]hat people do religiously in cities is shaped by what kind of cities they find themselves in, at what moment in the histories of those cities, and by their own experiences, cultural traditions, and contemporary circumstances."

In Berlin, as in other metropolises, globalization and migration have not only widened the range of religious choices and possible religiosities (Amin 2002; Eade 1997; Sassen 1999; Soysal 1994; Kuppinger 2015), but also brought along urban transformations with local forms and histories. While some religious traditions are naturalized as part of the nation-state's identity as a result of the national population's confessional past, other religious traditions are interpreted as foreign, suspicious, or unwanted.

This chapter examines the spiritualizing of Berlin through a discussion of how the multicultural image of Berlin becomes part of how Muslim youth create a "fun Islam," and the roles of mosques in this process. Young Muslims actively constitute the urban landscape as part of a long journey towards recognition of a minority religion, facilitated by the existing multi-ethnic and multi-religious urban culture which they, through this process, also contribute to shaping. The chapter first discusses how Muslims, participating in the religious youth organization Muslim Youth Germany (*Muslimische Jugend in Deutschland*, henceforth MJD), are involved in spiritualizing the urban landscape through their visualized religious expressions and practices. As a "multicultural" urbanscape, Berlin with its particular characteristics plays a part in how MJD participants negotiate their religious presence, at the same time as their religious practices enforce their local identity as Berliner Muslims.

Secondly, the chapter examines how MJD participants navigate among mosques and Islamic prayer spaces as they seek a religious everyday life in Berlin. Finally, mosques are examined not merely as visible manifestations of the Islamic presence in the city by virtue of being a religious building, but also through the multiple uses of the space. I analyze how the "Open Mosque Day" annual event constructs particular encounters between non-Muslims and Muslims. The chapter draws on my fieldwork with young Muslims participating in MJD and fieldwork at several mosques in Berlin in the periods 2004–07 and 2010–11.

The visibility of Islam in Berlin

Walking through the Berlin neighborhoods of Wedding, Kreuzberg and Neukölln, one encounters so-called religious businesses such as *Halal* butchers, cafes and restaurants, at every corner. The doner kebab is well established in the German fast-food market and is often cited as a positive effect of the Turkish presence in Berlin.[1] The multi-religious and multi-ethnic district of Kreuzberg has gained the image of being "multicultural," with the weekly "Turkish market" on Maybach-Ufer street on Tuesdays and Fridays. The market has to a large extent become a part of the city image, with several reviews on TripAdvisor. Apparently, the assortment and prices of headscarves at the Turkish market, which caters to local residents (both Muslims and non-Muslims) as well as tourists, can compete with those found in Istanbul. The market and its surroundings also attract people seeking to consume its "exotic" flair. Tourist busses wind through the crowded streets in the heart of Kreuzberg, introducing tourists to the local residents and their "ethnic" lifestyles. The increase in the number of young women donning the veil has resulted in the introduction of women-only hairdressers in these districts—with windows covered against the glances of passers-by and with men prohibited from entering. Wedding dress shops offer "Islamically correct" white dresses with matching veils. Marriage and henna parties (engagement parties) divided by gender are held in otherwise semi-public spaces by covering windows from the gaze of outsiders, thus enabling women to dance without headscarves. German public swimming pools offer "women only" hours, frequently used by young veiled women and sometimes their mothers. Young well-educated Muslims have established Internet platforms for Muslims in Berlin that gather information on religious, social and political topics, and present ongoing projects in Berlin. Other educated women are establishing Muslim day-care centers or work as teachers in Islamic schools. These creative inventions and initiatives indicate how, as Göle has noted, "Islam carves out a public space of its own, embodied in new Islamic language styles, corporeal rituals, and spatial practices" (Göle 2006: 6). In European cities, Muslim public spaces are also a consequence of social and even market forces through which Muslims seek visibility and legitimacy in the national public sphere (ibid.), creating "Muslim space[s]" (Metcalf 1996: 2).

Inner-city multi-ethnic neighborhoods are regularly portrayed as problematic spaces, reflecting social, economic, educational and employment inequalities. Yet multi-ethnic neighborhoods do not only harbor problems; they also participate in dynamic local cultural production that challenges cultural hegemonies. Small local shops provide opportunities for migrant entrepreneurship and are part of everyday life for local residents. These shops contribute to the production of unique local constellations and cultural innovations, although these innovations are not always recognized as such (Kuppinger 2014). Scholars of urban cultures (Zukin 2010) and religions (Orsi 1999) have drawn attention to the "creative cultural powers of multi-ethnic and multi-religious working-class quarters" (Kuppinger 2015: 199).

Urban cultures are also distinctly shaped by religious buildings, such as mosques, prayer rooms and religious organizations. In Berlin, the number of these spaces has risen steadily since the late 1980s: in 1999, there were about seventy mosques and prayer rooms in Berlin; in 2006, there were about eighty mosques and prayer rooms (Spielhaus and Färber 2006), and ten years later there are about a hundred (Spielhaus et al. forthcoming), a few of which are built with minarets. Since the middle of the 1980s, initiatives to construct purpose-built mosques with minarets have increased (Spielhaus and Färber 2006), frequently drawing significant media attention and public conflict (Cesari 2005; Jonker 2005; Oosterbaan 2014). The majority of these Islamic religious spaces are situated in the neighborhoods of Kreuzberg, Neukölln and Wedding, which are areas with a high percentage of residents with migrant backgrounds. Some mosques are located in the middle of the city: for instance, the Omar Ibn Al-Khattab Mosque, built in 2008 in Kreuzberg, is surrounded by bars, cafés and restaurants. Other mosques, such as the Şehitlik Mosque at Columbiadamm, are built in non-residential areas detached from the bustling city life. Furthermore, mosques are increasingly being built in the urbanesque parts of East Berlin (such as the Khadija Mosque in Berlin-Heinersdorf discussed by Hüwelmeier in this volume), where previously few Muslims visited, let alone lived. This expansion of mosques to East Berlin, which is one of the urban peripheries of Berlin that for long was avoided by Muslims from a fear of neo-Nazi groups, has been met with protests by the local non-Muslim population. This developing geography of the location of mosques in Berlin seems to take the shape of a circle: while at first mosques were mainly built in the inner parts city, today they are spread out in the urban and urbanesque areas of Berlin.

Yet mosques and prayer spaces in Berlin are typically located in run-down inner courtyards or old factory buildings. Sometimes a mosque is located in close proximity to other Islamic spaces; other times it is located by itself on a street where one would not imagine there to be a mosque unless one had inside knowledge. The entrances to these religious spaces are usually indicated by a sign (sometimes small, sometimes large) on the sidewalk directing visitors to a courtyard where the main entrance can be found. The shoe rack just inside the door marks the entrance to the place of prayer.

While around 90 percent of the mosques adhere to Sunni Islam, including some that are Sufi-oriented, there are also Alevite, Shiite and Ahmadiyya organizations (Spielhaus and Färber 2006). Some mosque and religious spaces are frequented by Muslims of a particular ethnic or national background (Turkish, Bosnian, Albanian, Indonesian, or Egyptian), which will be reflected in the language used in that particular mosque. Other mosques are attended by Muslims of different backgrounds who adhere to a particular Islamic religious orientation: orthodox, conservative, moderate, spiritual, or followers of a specific direction such as *Naqshbandi*—a Sunni spiritual order of Sufism. Still other religious spaces are popular among Muslims who wish to pursue religious activities and seminars in German. Mosques not only provide religious services, but also serve community functions, offering a place to be with others with similar spiritual approaches.

The public manifestations of mosques with or without minarets transforming the Berlin cityscape signal not only that Islam is there to stay, but also Muslims' desires to carve out spaces for Islam, in its various forms, in Germany. The religious, social and political roles of Islamic communities and organizations have changed over time, not only as Muslims' "migrant" status has changed but also as a consequence of the different social and religious needs of the new generation born in Europe (Bendixsen 2013a; Yükleyen 2007). Muslim youth born in Germany are increasingly creating, changing and molding Muslim spaces in Berlin, not only by establishing new organizations and fora for religious practices, but also by transforming existing Muslim organizations, for example, by initiating new public roles for mosques. From the mid-1990s in particular, the younger generation born in Germany has demanded institutional and administrative transformations within their parents' religious structures and organizations established in the 1980s. Through generational negotiations in the religious organizations, more space has been given to youth and to women in several of the established mosques and religious organizations, including through the creation of computer courses, sports clubs and separate women's activity spaces (Spielhaus and Färber 2006). Yet not all young Muslims feel at ease within their parents' organizational structures, which largely remain separated along ethnic and national lines (with the language of instruction being Arabic, Turkish, or Bosnian) and which frequently have a hierarchical structure.

I now turn to a discussion of one of the new forms of religious organizations that have been set up by a generation born in Berlin: Muslim Youth Germany (MJD). MJD was purposefully established with a different structure, ways of teaching Islam, and ways of establishing authority than the organizations founded by their parents. MJD represents one example of the agency and the resilience of Berlin.

Creating new religious spaces

MJD was established in 1994 under *Haus des Islam* (HDI, based in Lützelbach) and was founded by eight young Muslims aged between 17 and 20.[2] During my fieldwork (2004–07), MJD was located in an office behind the Muslim bookshop "Green Palace" and next to the organization Islamic Relief, along an otherwise deserted street in Kreuzberg.[3] Comprising approximately fifty branches Germany-wide, today it is the second largest Muslim youth organization in Germany, after the *Islamische Gemeinschaft Milli Görüs* (IGMG). MJD Berlin has a "sister" and a "brother" group that meet on different days of the week. MJD can be understood as an urban phenomenon: Fatima (31) told me that during a Muslim camp in 1994, a Muslim "brother" from the United Kingdom talked about an organization for Muslim youth in England. Fatima remembered sitting around a camp fire saying, "would it not be great if, living in Hamburg but going to Aachen, you would know where you could find other young Muslims who are actively practicing their religion?" She added that one of the purposes of MJD is to offer a place for youth to be "Muslim and German, but also cool." At the local level, the organization targets Muslim youth of both genders who wish to live out an "authenticated Islam" (Deeb 2006), striving to be or become devout Muslims in Germany. MJD seeks to attract young so-called "born" Muslims who are religiously "inactive," and on a national level it takes on a public role in representing Islam in Germany by initiating and responding to invitations to dialogue with politicians, journalists and interfaith projects (Bendixsen 2013a).

MJD is organized by youth for youth (aged between 15 and 30) and is detached from any particular mosque. In contrast to the mainly ethno-homogenous mosques and organizations in Germany, most participants in MJD in Berlin have been born or raised in Germany and have parents from different nationalities or ethnic groups, including Bosnia, Egypt, Turkey, Palestine, Syria, Sudan and

Figure 8.1 The street in Kreuzberg (city district of Berlin) where Muslim Youth Germany (MJD) was located during the fieldwork. Photo by Synnøve Bendixsen.

Kurds from Turkey and Iraq. The group is frequented by mixed-marriage children (for example, those with a German, frequently converted, mother and Egyptian father) and converted Germans.

During the weekly meetings, I could observe how the youth educated themselves by reading from the Quran in Arabic and then in German, "so that they can understand what they have read," followed by a *tefsir* study (Quran commentary or exegesis) prepared by one of the young women. The agenda continued with presentations on how to be a "good" Muslim or on issues relevant to everyday life as a young Muslim in a non-Islamic society such as Germany. The youth distanced themselves from their parents' religious practices, which they viewed as traditional and backward, informed by their parent's national or ethnic background and not necessarily Islamic. Instead, MJD youth sought a "universal Islam" and actively detached it from its Turkish or Arabic traditions. Youth sought to become "good Muslims" by combining religiously informed practices with a youthful and "fun Islam," that aimed to bridge being German, Muslim and young. As one of the older participants told me, MJD was a place for "young Muslims to not feel that they are stupid, but that they are normal and that faith can also be fun, without segregating themselves from society."[4] MJD participants are focused on being integrated in German society, have relatively high levels of education, carry out all activities in German, and seek a conservative Islam combined with "pious fun" activities in the urban realm; this detaches them from their ethnic enclaves in the city and brings them together as German Muslims. Berlin stages the opportunity to pursue "fun Islam" and being pious or devout young Muslims, making MJD very much an urban phenomenon.

To experience practices in an "Islamic ambiance" (Bowen 2010) is not only restricted to the mosque or prayer space; any urban space can be spiritualized depending on the activity involved and communal orientation. In addition to the weekly events, some MJD youth are in charge of organizing larger events (for instance, a hip-hop concert with Muslim convert Ammar 114) and different leisure activities. On Fridays or Saturdays, the young women visit each other at home where they dance, tell jokes, listen to world music, sing karaoke, play different games, and spend the night. The young women often go bowling together, enjoy ice-skating in Wedding, go to the movies, swim during particular hours for women, and arrange picnics in different city parks all over Berlin in order to "get to know Berlin better. After all, this is where we live." During these events, they explore new parts of Berlin, pray together (sometimes in a park), and discuss religious themes such as how to combine being a "good Muslim" and a "good citizen" in Germany.

This generation of well-educated and socially active youth combines a desire to be pious Muslims, modern, and having fun without seeing these as contradictions (Bendixsen 2013b). This is possible in an urban atmosphere such as Berlin, where they can choose between the various spiritual approaches and religious spaces. More specifically, to build on Orsi's (1999) analysis of the relation between religiosity and the urban environment: what role does the city of Berlin play in what these youth do religiously?

Becoming a multicultural Muslim in Berlin

Devout Muslim youth are continuously confronted with questions such as "are you Muslim *or* German?" and "where do I belong?" (Bendixsen 2013b; Kuppinger 2015). By transcending such distressing questions, MJD represents a space where these youth can create a lifestyle of being devout Muslim Germans. Being multicultural is part of this self-representation. The fact that they all have different backgrounds (and are not only of Turkish descent as is common in other religious spaces) is frequently emphasized as a positive marker of the group. During MJD events, in online descriptions of the organization, and in discussions among the youth, the importance of challenging the general presumption (made by their parents, teachers and non-Muslim surroundings) that it is impossible to be both German and Muslim is emphasized. MJD's multicultural image ties its practices to Berlin: Berlin is (in)famous for being multicultural, a characteristic that is frequently used positively to attract tourists, but which is either promoted or discredited when media and politicians discuss migrants' loyalty (or lack thereof) to Germany, seek to explain social problems, and/or look for "proof of a politics gone wrong."

In explaining why she considers Berlin her home and the only place she could live in Germany, Nuriye, a 25-year-old woman with Turkish parents, said, "It's multicultural, like nowhere else in Germany!" The multicultural aspect of MJD located participants' Muslim-ness as part of and belonging in Berlin, that is, being multicultural and Muslims make them Berliners. They become owners, participants and shapers of Muslim tradition and the local cityscape (Henkel 2007). But Berlin is more than multicultural: it is multi-ethnic and multi-religious, embracing, and continuously produced by, a diversity of Muslim communities and Islamic orientation.

Indeed, moving to Berlin with their parents had explicit consequences for the religiosity, religious practice and experiences of some youth. Aischa, a 21-year-old woman, grew up in Germany with parents who migrated from Turkey. Born in a small town in West Germany, she moved with her family to Berlin four years before I first met her. One sunny spring day, we had just bought ice cream and were walking down a crowded shopping street in Wedding close to where she lived. Looking around her, she said:

> There are so many Turks here [in Berlin], but only very few who you would want to talk to. Look at them, [without a thought in their heads]. After such a long time without really having anyone I am interested in spending time with, it so good to have found MJD, like, people you can really talk to.

Her 17-year-old sister explained that what she liked about Berlin was that she could find so many different groups, "that there are spaces for everyone."

Likewise, Ines had grown up in a small German town where she seldom socialized with people with migrant backgrounds. She had mostly "ethnic" German friends before the age of 14. After a turbulent youth, she became more religious and decided to veil at the age of 17. She made this decision after she moved to Berlin:

> And then when we moved here [to Berlin], it was . . . the summer of 2002, I saw foreigners, I was shocked. It was a different mentality. I saw people go to the mosque and pray there. It was a whole new world for me here. And I started to learn more about my religion, and the more I read the greater was my [desire] to wear the headscarf. Here they think that you are "easy to get," even the Turkish men think that. I wanted to show that I am a Muslim. To limit, not like limit so that I don't belong here, but to show that I am a Muslim.

Moving to Berlin became a turning point for Ines in that she encountered practicing Muslims and became more familiar with Islam. Exposed to a variety of lifestyles through her earlier experiences in a children's home where she had lived for a while, she now started to dismiss "all the freedom" she had been given and viewed it as superfluous and not what she required. The veil, she thought, would create the necessary gendered distance in the public sphere.

Both Aischa's and Ines's stories illustrate how living in Berlin can shape religious participation, spirituality, and how to lead a religiously informed life. For both women, Berlin as a city offered a variety of religious spaces representing different religious orientations and congregations providing particular infrastructures for action. Ines's "turn to Islam" within the Berlin urban space can be understood as shaped by her experience of others' practicing Islam differently, but also a fear of how others would view her. While for Aischa, the pool of various religious groups facilitated her search for a space in which she was comfortable living out her religiosity.

For youth participating in MJD, its multi-ethnic composition and multicultural image shaped how they performed their religious practices and how they viewed themselves and the organization as part of Berlin. In the search for an "authenticated Islam" (Deeb 2006), these youth distance themselves both from their parents' practice of Islam, which they view as traditional, and from what they consider a secular society. This distancing involved learning about Islam "the right way," which meant understanding how a particular religious practice or obligation should be performed and for which reasons. Their practice of Islam also meant wearing the headscarf in "a proper way": that is, covering the ears and neck and not using too strong colors or the grid or flower styles of their "traditional" mothers; many of whom still wore the type of headscarves (shape, color and texture) they had worn in their villages before arriving in Germany. Veiling fashion is part of the construction of a religious self as youthful and urban; and inspiration for how to tie the veil, the use of ornaments and color was found on the Internet, in series on satellite TV (in particular, Turkish soaps), and through discussions with friends (Bendixsen 2013c). The young shoppers bought their headscarves at the Turkish bazaar, in shops selling Islamic fashion in Kreuzberg, Wedding, or Neukölln, or while visiting family in their parents' countries of origin. But headscarves also came from European chains stores such as H&M and C&A, located on the popular shopping street Kurfürstendamm. Stylish clothes from these shops were made "Islamically correct," which means covering the body and not being "too sexy," by buying them one size too large and layering. In a discussion in the youth group about what to wear, one youth jokingly said, "We have to be modern, after all we are Muslims!" While making fun of how they perceived themselves to be viewed as "non-modern" by non-Muslim strangers, this comment also reflects how the youth view themselves: they *are* religiously devout and modern youth.

A couple of years after I had ended my intensive fieldwork with MJD, I meet Ines by chance standing in line in supermarket in Kreuzberg. She was now wearing a full face-veil, a decision she had taken after getting married, although against the wishes of her husband who was of Lebanese descent. The more she had read about duties in Islam, the more she had realized that this was more correct than what she had learned in MJD. I also knew that for a while she had attended the notorious Salafi al-Nur mosque in Neukölln, where other young women had started wearing the face-veil. While she remained convinced that face-veiling made her closer to God and Paradise in the afterlife, she also acknowledged that this Islamic style ultimately had changed her everyday life in the streets of Berlin. She regularly experienced that non-Muslim strangers stared at her with discontent and some even shouted at her in the street. Ines preferred it when strangers openly asked her about her Islamic practice, but sometimes such conversations led to openly hostile quarrels. I realized that, to Ines, her religiously informed decision to wear the face-veil fundamentally changed her experiences of the city, of the way she moved in the urban public sphere, and of how to be a Muslim in Berlin.

Modified desires can bring along alterations of dress and thus render one's religious identification visible. A full face-veil or veiling in Berlin public space means that one is visible. Public space is a field of gazes and a site of social recognition or rejection. It is also an arena of communication and potential (mis)understandings. Further, Ines's story alludes to how religious practices and desires are not fixed but can be reworked, motivated by the religious space one attends, the literature one reads, and the religious community in which one finds themselves belonging.

Urban people present themselves and are observed as specific moral agents, but also as makers of cultural worlds in the street (Orsi 1999). Ines's decision to start veiling after moving to Berlin is indicative of the place of the profane in the city and the sociability of the urban streets. Because there are so many different kinds of desires in the city, style becomes particularly important to urban people and to urban religion. As Orsi argues:

> It is through style—through the intricate intentionality of public self—representation, and especially through style in religion—that city people have made meanings and impressed those meanings on themselves and others, have met and contested the gaze of reformers and bureaucrats, and have presented themselves at the borders and junctures of adjacent urban social worlds. (Orsi 1999: 49)

Religious styles can be altered and shaped in conjunction with personal desires, family expectations and friends' fashion, but also by the religious community space and its spirituality. Among all these different religious spaces and desires in urban and urbanesque space, how are the youth navigating which of these to attend?

Navigating among religious spaces

Several of the youth in MJD had like Ines attended the al-Nur mosque, which is located in an old factory in a run-down industrial area—an urbanesque habitat—in Neukölln. One of the reasons why youth from all over Berlin visited this mosque, even though some of them had to change trains more than twice to get there, was the presence of Abdul Adhim Kamouss, a charismatic imam of Moroccan origin. During my fieldwork, I observed many young people of both genders walking from the train station to the mosque during the weekends. Some Sundays, there would be more than a hundred women seated in the women's section, waiting for the TV broadcast of Kamouss's talk, which he delivered downstairs in the men's section.

When MJD was established in 1994, in many ways, it filled a gap for the younger generation by offering a space for the creation and exchange of spiritual and religious ideas and practices. Today, there are several other Muslim spaces targeting youth and there is a certain competition for membership and attendance among the religious organizations in the city. Greater choice also means

Figure 8.2 Al-Nur mosque in the urbanesque part of Neukölln (city district of Berlin). Photo by Synnøve Bendixsen.

an increase in changes of allegiance and attendance, as youth change between different religious communities because they do not agree with the religious organization's cause or teaching method (Yükleyen 2007), or because they are dissatisfied with the social interaction with the other participants. There is a diversification of Islam in Berlin, with mosques and religious groups adhering to and practicing different forms of religiosity from perspectives that could variously be called spiritual, universally oriented, national and ethnic oriented, moderate, conservative and fundamentalist (these are not mutually exclusive). Youth seeking to take active part in a religious community are frequently mindful of where to go, sometimes trying out various places, sometimes following the advice of friends or parents, and sometimes merely visiting the organization or mosque located in their neighborhood.

While some youth had a specific mosque they always attended, others said "I never went to only one mosque. Many people are going to one particular mosque, like to a Turkish or Bosnian mosque and so on, and belong to that. I don't do that. I go everywhere." I frequently accompanied the youth as they explored yet another mosque in a different neighborhood, either because they were curious about the place, wanted to hear the imam there, or because the mosque was nearby when they wanted to pray. Some mosques typically attracted mainly Muslim residents in that particular neighborhood. Still other mosques, such as the al-Nur mosque in Neukölln, were particularly popular among youth in Berlin at certain moments in time. The popularity comes from a charismatic imam, from the mosque arranging activities that attract youth or because the mosque represents a large social network, which was essential in the search for a devout spouse. There were other mosques that the youth avoided because they had a negative reputation ("they are on the list of extremists" or "the women are too strict there"). Aishegül, a 29-year-old university student who grew up in Neukölln, where her parents had settled when they migrated from Turkey in the 1970s, talked about how she decided which mosque to attend:

> During Ramadan, we pray together, you know, and then I usually go to the one close to where I am. Sometimes I go to an Arabic mosque. They recite very nicely, read the Koran beautifully. But there are always so many children there, and they make so much noise that sometimes you cannot even hear what they say. That is why sometimes I go to the Turkish mosque. The Turkish are really strict. And they have fewer children with them . . . And sometimes I go to the Bosnian mosque . . . they read so beautifully, some are really good. That is, all are good, but some are even better. And you would like to see who is reading it. There [in the Bosnian mosque] you can see the men, because the women and men are not divided. They have only one room, that's why. And also they are not that many, so it is possible.

She added: "You go to the mosque that is there when you need it." Because she wanted to perform the five obligatory prayers each day, she had found a solution when a mosque was not nearby during the prayer slot (the period of time within which a prayer must be held): "if you are not at your place or at an appropriate place when you need to pray, then I always go to a dressing room, for instance at H&M."

Aishegül's comments suggest that while mosques and religious organizations are significant in religious everyday practices and in creating a sense of belonging for Muslims, Muslims give different meanings to these religious spaces and develop different uses for and attachments to them. Indeed, cities are the stage and accommodating space for diverse religious ontologies (Orsi 1999). In order to better understand how religious organizations and their participants take and make spaces (Knott 2005) through public visibility, I now turn to how mosques also become embedded in urban cultural negotiations and transform the urban space. I do this by examining the annual "Open Mosque Day" event.

Repositioning mosques in the urban landscape: "Open Mosque Day"

There is a constant (negative) focus on what is going on in the mosque, with documentaries, politicians and the media questioning the attitudes conveyed there. Scandals, such as when an imam's hate speech or belittling of non-Muslim Germans has been recorded or when it emerges that a mosque is on the watch list of the domestic intelligence service of the Federal Republic of Germany (*Bundesamt für Verfassungsschutz*, BfV), contribute to mosques becoming sites of fear, suspicion and anxiety in the German political and public sphere. In the dominant public sphere, mosques are viewed as spaces where youth can be persuaded to withdraw from dominant German society, or at worst be turned into terrorists. Fekete (2004) argues that the German system of religious profiling of foreign nationals from Islamic states is unprecedented in scale. German politicians have proposed making mosques more transparent by requiring imams to preach in German and by monitoring all mosques by closed-circuit TV cameras.

Since the late 1990s, Muslims actively engaged in mosques have invited non-Muslims to visit mosques by organizing special events. Such efforts include developing closer connections with organizations in the neighborhood of the mosques (Kapphan 1999), inviting school classes for tours in the mosque, and establishing the "Tag der offenen Moschee" (TOM, "Open Mosque Day"), the latter of which I will discuss further. Since 1997, October 3 has been Open Mosque Day in Germany. This day was initiated by *Zentralrat der Muslime* (ZDM, Council of Muslims in Germany), and since 2007, it has been organized by *Koordinationsrat der Muslime* (KRM).[5] On this day, several hundred mosques invite the German-majority population to visit and offer a special program in the mosques for non-Muslims. More than 100,000 people across Germany visit a mosque on Open Mosque Day, according to KRM.[6] In Berlin, 15–18 mosques participate every year, each belonging to one of the three major Islamic umbrella organizations: the Turkish Islamic Union of Establishments for Religion (DITIB), the Islamic Federation (IFB), and the Berlin Association of Islamic Cultural Centers.[7] Each mosque has anywhere from 30 to 4,000 visitors; this figure has been more or less constant for the last few years.[8] Most of the mosques that participate in the event offer mosque tours, panel discussions, bookstalls and folklore performances. A special TOM working group (ZDM) provides materials for the mosques to use, including advertising material, posters, handouts and fliers (on topics such as "What is a Mosque" and "What is Islam"). In one of the mosques in which I participated, they also invited non-Muslims to participate in one of the five daily prayers.

Open Mosque Day in Berlin is a creative praxis to make mosques, in the abstract and physical sense, more visible in the city landscape. This cultural event makes Islam visible and produces Muslim public space in a different way: at the city and neighborhood level, non-Muslims can find out about the location of mosques and what goes on there without having to rely on media (mis)representations. While cultural and religious practices such as the Open Mosque Day can work to oppose a negative meta-narrative about Islam and mosques as "a hideout for terrorists," as I will show, this day also produces new cultural practices in the mosques, if only for one day. I illustrate this by comparing the use of space on October 3, and on regular days in three mosques (two in Wedding and one in Neukölln) in which I did fieldwork every Friday and on Open Mosque Day (2004 and 2005).

The German speaking El Nur Mosque is located on the ground floor in a courtyard building in Wedding.[9] There are two entrances, one for women and one for men. The main room where presentations and Friday prayers take place is quite small. The women and men are divided by a thin mobile wall so that the men and women cannot see each other even though they are in the same room. The imam and lecturer sit at the end of the mobile wall, in the middle of the room, so that he or she can be seen by both sides. After the presentation, the men use this room for socializing, while the women withdraw to a room between this main room and the kitchen. On regular Fridays, children circle between the two gendered spaces, to get more cake or tea for the men, to pass messages between the women and men, or for their own amusement.

In contrast, during Open Mosque Day, gender mixing was accepted, not only for the non-Muslim visitors, but also among the Muslims. Tables and benches were set up for this occasion (though regular mosque participants normally sit on the floor), and men and women ate together in the kitchen area. Inside the main prayer room, the dividing wall usually separating the genders had been removed, and a bookstall from the German Muslim bookshop "Green Palace" had been set up. In each of the three mosques, the organizers had installed Oriental images and decorating styles for that particular event: in one mosque, large colorful Turkish or Arabic pillows were placed in a corner where tea could be enjoyed from elegant teapots usually associated with the Orient, and specialty food from Muslim countries was served instead of the biscuits from the nearby Lidl's supermarket served on a regular day. Visitors were asked to remove their shoes before entering the mosque, but unlike when visiting mosques in parts of Asia or the Middle East, visitors were not asked to dress modestly on this warm summer day.

The choice of October 3 as Open Mosque Day is no coincidence: it is the day of German unification and a public holiday in Germany. According to the organizers, the day was chosen because "it should express the self-reflection of Muslims as part of the German unification and their solidarity with the general population."[10] Holding Open Mosque Day on this date also implicitly challenges what is "German" and makes a claim for symbolic space within national culture. The motto and main topics the Central Council has chosen in recent years indicate that this day is valuable as a means of dialogue and transparency. In the early years of Open Mosque Day, the mottos for the day included "Islam and Violence," "Muslims and Internal Security," "The Headscarf Debate," and "Muslims: Partners Against Racism." In recent years, the mottos have been "Environmental Protection—Mosques Are Committed" (2013), "Social Responsibility—Muslims For Society" (2014) and "Young Muslims In Germany—Motivated, Committed, Active" (2015).

During the open day, the mosque was turned into something other than a space for devoutness: it became a performative space where Islam was exhibited but not practiced. In this performance, mosques and Islam are made into something harmless, exotic and open. This act of transformance for one day is important in the effort to make being Muslim a regular part of everyday Germany. The event is part of an effort to redefine mosques as public spaces, signaling a willingness to communicate outside the Muslim community and to counter non-Muslims' fears about "what is going on" within the walls of the mosque in their neighborhood. It positions the mosque in the urban landscape and in a larger cultural sphere. The local non-Muslim population is given the opportunity to meet Muslims and Islam face-to-face rather than through the media and political discussions. As the organizers put it:

> Muslims will do their best to answer questions during Open Mosque Day and be available for discussions. It is perhaps not so much about having a good command of the German language, but more about the wish to interact in good ways.[11]

As such, the mosque becomes an alternative public sphere where Muslims can be more in control of the premise for the encounter between Muslims and non-Muslims. Organizers explicitly intend to detach the concept of the "mosque" from associations with "terror" and to replace this with a positive experience. The statement by Dr. Nadeem Elyas (ZDM Chairman) for TOM 2004 is indicative of this: "In some places, mosques and Christian communities and Muslims and non-Muslim neighbors lived next to each other for decades, without either one daring to knock on the other's door and walk across the stranger's threshold."[12]

Open Mosque Day becomes a platform for discussions about what it means to be a German Muslim. As a cultural practice, it creates connections between religious practice, the local neighborhood, the city and the dominant society. It creates space for not only imagining but also enacting the role of Islam and Muslims in Berlin. Perhaps paradoxically, the opening of the mosque to non-Muslims also involved the transformation of the rooms and a change in the regular practices within that space. In this process, the aesthetics of Islam is changed and in some ways come to resemble more that of an Orientalist style of Islam. The preparation for receiving a large number of visitors in the mosque included reconstructing the space to be accessible for non-Muslims. In all three mosques, the way the rooms were used and the genders were divided changed for this day. The space was transformed in that gender mixing and the dress habits of the dominant German society were allowed instead of the usual gender segregation and religious dress code.

Simultaneously, the event normalizes the mosques within the complex and creative urban religion and culture (see also Kuppinger 2014). The event inverts the regular order: in Germany, many Muslims experience that the non-Muslim majority society has definitional power over what Muslims and Islam are thought to be. However, during Open Mosque Day, the marginalized invite the dominant society to a dialogue and are in charge of the agenda and arena. Many mosques invited local authorities, such as the local police and neighborhood organizations, which had a positive impact on the relationship between the mosque and the local community. Signs and posters on the street and the mixture of people outside the entrances made the mosque more visible than normally. This annual event challenges the stereotypical image of Islam among non-Muslims, and contests what and who has the right to be visible in the urban landscape. It is embedded in the need to construct open and porous forms of being in public as a way of "living together with strangers in the present" (Morley 2001: 441). Within this framework, the mosques become sites of cultural exchange and transformation. Open Mosque Day transforms the mosque into a semi-public space where dialogue is facilitated on a stage set by the minority. This allows mosque members to regain control of the meaning these spaces have in order to make them the appropriate representations of a stigmatized group.

Conclusion

Berlin is described—sometimes positively, other times negatively—as a multicultural city with a high percentage of migrants and characteristic inner-city streetscapes such as the main street of Kreuzberg, Oranienstrasse, where a gay bar and a mosque frequented by Turkish ultra-nationalists are located next to sushi restaurants, cocktail bars and a Turkish bakery. The vast number of mosques in Berlin, scattered across different locations in the urban and urbanesque environment, not only attests to spirituality in the city but also allows for different ways of being religious. As Orsi has pointed out:

> Religious maps constitute as they disclose to practitioners particular ways of being in the world, of approaching the invisible beings who along with family members and neighbors make up practitioners' relevant social worlds, and of coordinating an individual's own story with an embracing cultural narrative. (Orsi 1999, 53)

Just as young Muslims actively make use of urban spaces in their religious practices, opening the mosque for a day contributes to a reclaiming of control of the meaning of Islam and of being Muslim in twenty-first-century, multi-religious

Berlin. Open Mosque Day can be understood as an effective rhythmic cultural event that aims to exert social influence over the image of mosques in Germany. It offers an arena for dialogue that facilitates encounters between people with different religious or secular lifestyles. Such practices of spiritualizing the city is not "the antithesis of the city," but rather an integrated urban form of living that contributes to the resilience of cities.

Being "multicultural Muslim youth," MJD participants articulate their Muslim present as a part of Berlin's well-known multicultural characteristic. They are agents of a "transformative project of spatial appropriation and re-enchantment of the urban landscape" (Garbin 2013). While the urban, through its manifestations of consumption, diversity, and alternative spaces, is generally considered to offer and even tempt one to a liberal lifestyle, the urban can also provide the opportunity and desire to become more religiously conservative or orthodox, or provide the grounds for a new form of spiritualization. Multicultural Berlin is also multireligious, fostering and offering different ways of being and becoming a Muslim youth. "Fun Islam" is an innovative practice that these youth engage in, informed by their spiritualization and the urban sphere in which they live. Pursuing modern veiling and devout fashion and transforming urban spaces into religious sites, the youth are both accentuating their differences and claiming rights to be young, religiously observant, and German—in other words, to be Muslim Berliners.

Notes

1 In contrast to the majority of ethnic businesses, doner kebab shops targeted Germans as consumers from the beginning, and about 95 percent of doner customers are ethnic Germans (Caglar 1995).

2 I conducted intensive fieldwork with this organization (2004–07), which was followed by more random meetings and conversations. I also interviewed its participants in 2010.

3 When the bookshop went out of business in 2010, MJD moved its meetings to different places.

4 Julia Gerlach (2006) calls participants in MJD and this generation of well-educated and socially active youth "pop Muslims," a term I object to for two main reasons: first, during my fieldwork, many of the youth argued that the book's terminology respected neither them nor their activities; second, the term not only neglects, but also takes attention away from, how the religiosity of the youth is part of a complex process where the youth actively interact with religious sources.

5 Established in 1994, the Council of Muslims in Germany is an umbrella organization for more than 19 Muslim societies. It supervises about 500 mosques.

6 http://tom.igmg.org/index.php; accessed February 15, 2016

7 An incomplete list of the mosques participating in the event can be found online at www.islam.de.

8 ZDM website. Although ZDM claims on its website that this concept is unique in the world, there are examples of open-door mosques, including in southern California (since 2004), the UK (organized by the UK Islamic Mission since 2002) and Norway.

9 Regular attendees are Egyptians, Turkish, Palestinian, Iraqi Kurds and converted Germans and Polish women.

10 http://www.islam.de/2583.php; accessed January 9, 2005.

11 www.islam.de; accessed December 19, 2006.

12 www.islam.de; accessed December 19, 2006.

References

Amin, Ash (2002), Ethnicity and the Multicultural City: Living with diversity, *Environment and Planning A*, 34: 959–80.

Bendixsen, Synnøve (2013a), *The Religious Identity of Young Muslim Women in Berlin. An Ethnographic Study*. Leiden: Brill.

Bendixsen, Synnøve (2013b), Being Muslim *or* being 'German'? Islam as a New Urban Identity, in Wilcox, Melissa M. (ed.), *Religion in Today's World. Global Issues, Sociological Perspectives*. London: Routledge.

Bendixsen, Synnøve (2013c), 'I love my Prophet'. Religious Taste as Social Boundary Marking in Berlin, in Moors, Annelies and Tarlo, Emma (eds), *Islamic Fashion and Anti-fashion. New Perspectives from Europe and America*. London: Berg.

Bowen, John R. (2010), *Can Islam Be French? Pluralism and Pragmatism in a Secularist State*. Princeton, NJ: Princeton University Press.

Caglar, Ayse S. (1995), *McDöner: Döner Kebab* and the Social Positioning Struggle of German Turks, in J. Costa and G. Bamossy (eds), *Marketing in a Multicultural World*. Thousand Oaks, CA: Sage.

Cesari, Jocelyne (2005), Mosque Conflicts in European Cities: Introduction, *Journal of Ethnic and Migration Studies*, 31(6): 1015–24.

Deeb, Laura (2006), *An Enchanted Modern*. Princeton, NJ: Princeton University Press.

Eade, John (1997), Reconstructing Places: Changing Images of Locality in Docklands and Spitalfields, in John Eade (ed.), *Living the Global City. Globalization as a Local Process*. London and New York: Routledge.

Fekete, Liz (2004), Anti-Muslim Racism and the European Security State, *Race & Class*, 46(1): 3–29.

Garbin, David (2013), The Visibility and Invisibility of Migrant Faith in the City: Diaspora Religion and the Politics of Emplacement of Afro-Christian Churches, *Journal of Ethnic and Migration Studies*, 39(5): 677–96.

Gerlach, Julia (2006), *Zwischen Pop und Dschihad. Muslimische Jugendliche in Deutschland* ["Between Pop and Dschihad: Young Muslims in Germany"]. Berlin: Ch. Links Verlag.

Göle, Nilufer (2006), Islamic Visibilities, in Ludwig Ammann and Nilufer Göle (eds), *Islam in Public. Turkey, Iran and Europe*. Istanbul: Bilgi University Press.

Henkel, Heiko (2007), The Location of Islam: Inhabiting Istanbul in a Muslim Way, *American Ethnologist*, 34(1): 57–70.

Jonker, Gerdien (2005), The Mevlana Mosque in Berlin-Kreuzberg: An Unsolved Conflict, *Journal of Ethnic and Migration Studies*, 31(6): 1067–81.

Kapphan, Andreas (1999), Zuwanderung von Muslimen und ethnischen Gemeindestrukturen, in Gerdien Jonker and Kapphan Andreas (eds), *Mosques and Islamic life in Berlin*. Berlin: Ausländerbeauftragte des Senats.

Knott, Kim (2005), *The Location of Religion: A Spatial Analysis*. London and Oakville, CT: Equinox.

Kuppinger, Petra (2014), Flexible Topographies: Muslim Spaces in a German Cityscape, *Social & Cultural Geography*, 15(6): 627–44.

Kuppinger, Petra (2015), Pools, Piety, and Participation: A Muslim Women's Sports Club and Urban Citizenship in Germany, *Journal of Muslim Minority Affairs*, 35(2): 264–79.

Metcalf, B. D. (ed.) (1996), *Making Muslim Space*. Berkeley: University of California Press.

Morley, David (2001), *Belongings. Place, Space and Identity in a Mediated World*. London: Sage.

Mühe, Nina and Riem Spielhaus (forthcoming), *Islamische Gebetsräume in Berlin*. Berlin: Der Beauftragte für Kirchen, Religions- und Weltanschauungsgemeinschaften.
Oosterbaan, Martijn (2014), Religious Experiences of Stasis and Mobility in Contemporary Europe: Undocumented Pentecostal Brazilians in Amsterdam and Barcelona, *Social & Cultural Geography*, 15(6): 664–82.
Orsi, Robert (ed.) (1999), *Gods in the City*. Bloomington: Indiana University Press.
Sassen, Saskia (1999), *Guests and Aliens*. New York: The New Press.
Soysal, Yasemin Nuhoğlu (1994), *Limits of Citizenship: Migrants and Postnational Membership in Europe*. Chicago, IL: University of Chicago.
Spielhaus, Riem and Alexa Färber (2006), *Islamisches Gemeindeleben in Berlin*. Berlin: Der Beauftragte des Senats für Integration und Migration.
Yükleyen, Ahmet (2007), "The European Market for Islam: Turkish Islamic Communities and Organizations in Germany and the Netherlands," unpublished PhD thesis, Boston University, Graduate School of Arts and Sciences.
Zukin, Sharon (2010), *Naked City. The Death and Life of Authentic Urban Places*. Oxford: Oxford University Press.

Internet sources

http://tom.igmg.org/index.php; accessed February 15, 2016.
http://www.islam.de/2583.php; accessed January 9, 2005.
www.islam.de; accessed December 19, 2006.

Part III

The agency of body and senses in spiritualized practices

9 The dance floor as urban altar

How ecstatic dancers transform the lived experience of cities

Sarah M. Pike

As I walk up from the Oakland subway on a Sunday morning, rain is falling in a slow drizzle and the downtown street is deserted. Around the corner I spot the "Tropicana Ballroom" sign on a 1920s building that is my destination. I pay $15 at the entrance and walk up the red-carpeted stairs to a ballroom. Art deco wall sconces softly glow onto an 8,000-square-foot floor full of dancers warming up. On the far side of the dance floor, beneath floor-to-ceiling windows, a vase of gladiolas, candles, a statue of Shiva and Zen Tarot cards have been placed on a table to create an altar. A young woman sits cross-legged in meditation in front of the altar; the man next to her is kneeling and praying, and two hundred other dancers—ranging in age from infants wearing padded ear coverings to men and women in their seventies—are preparing to "sweat their prayers" on the dance floor. In Los Angeles, Oakland, Dallas, New York, Amsterdam, Berlin, and many smaller cities and college towns, "Silicon Valley Shamanic disco," "Sacred Groove," "Barefoot Boogie," "Ecstatic Sunday Mass," "Dance Temple," "Mindful Meltdown," "Soul Sanctuary Dance" and "Soul Shake Dance Church" attract dancers of all ages and ethnic backgrounds for whom movement is a prayer and the dance floor an urban altar.[1] At urban dance events, ecstatic dancers transform the lived experience of cities at the same time that cities make possible these spaces for dancers' own spiritual transformation.

At weekly urban ecstatic dance gatherings across the contemporary spiritual landscape of the West, dancing bodies and dance churches are transforming the lived experience of cities in what the Californian dance teacher Shiva Rea describes as an "incredible revival of free-form movement."[2] At these dance events, organizers and participants attempt to shore up the boundary between the city and the dance sanctuary. They contrast an "open" self, made possible in the dance space, to the closed self outside. However, in practice, these boundaries are always breaking down. Through a variety of strategies of remembering and forgetting, dancers set apart their dance spaces from the layered history of their city environs, imagining them as free-floating and future-looking, purified of the street grit and complex histories outside that mark the urban topography. Described on one 5Rhythms® community's website as "an antidote to city life," dance worship has nevertheless been powerfully shaped by the urban context in which it developed. Dancers carry their urban lives with them into the dance and take their dance experiences out to city streets and into their homes and neighborhoods. Although dancers transform these spaces into temporary temples of movement, dance floors still carry traces of their other lives and demonstrate the complexities of attempts to separate the sacred from the city. Surrounded by bustling urban life, dancers come to the floor with heightened expectations that the dance will provide a contrast: a sanctuary or escape from the city outside, making ecstatic dance spaces more powerful *because* of their urban location rather than in spite of it.

School gymnasiums, yoga studios, fraternal organization halls such as Oddfellows and Sons of Italy, community centers, old warehouses, ballrooms in cities across the United States and Europe host the growing ecstatic dance movement. In the US, ecstatic dances are more than coastal phenomena in California and New York. They have spread to cities in the middle of the country like Dallas and Denver as well as college towns like Bloomington, Indiana where I attended "Trance Dance," a small gathering of ecstatic dancers in the fall of 2014. Towns like Bloomington exemplify the urbanesque nature of ecstatic dance, since outside of urban areas, it is usually found in college towns or suburbs inhabited by dancers who are likely to have been influenced by their own urban backgrounds or the teachings of dancers based in cities, like Gabrielle Roth, the founder of 5Rhythms® dance in New York. Many of these dance events are based on self-described "urban shaman" and Gabrielle Roth's 5Rhythms® therapeutic dance teachings. Other dance modalities developed in urban settings also shape dance gatherings across the country, such as Vinn Arjuna Marti's Portland-based Soul Motion™ Conscious Dance Practice that synthesizes Marti's dance training, New Thought Christianity, meditation and yoga. "Ecstatic Dance" as practiced in Oakland at Sweet's Ballroom was started in Hawaii by Max Fathom who had learned of Gabrielle Roth's 5Rhythms® and the Body Choir in Austin, Texas, and merged these techniques with DJ culture. Donna Carroll and Tyler Blank then brought Ecstatic Dance from Hawaii to Oakland in 2008.[3] But like Roth and Marti, they remain in the background of drop-in ecstatic dance events, even though they provide much of the inspiration (in the case of Roth) and organizational energy (in the case of Blank and Carroll) that makes the events happen. Much of the work of running dance events—stringing lights, setting up the sound system and creating altars—is done by volunteers.

Since the publication of Roth's *Sweat Your Prayers: The Five Rhythms of the Soul* in 1997, ecstatic dances have emerged in an urban context of raves and other electronic dance music events and also incorporate strands of the New Age movement. For Roth, soul-changing experiences are facilitated by dancing to five rhythms that she named "flowing," "staccato," "chaos," "lyrical" and "stillness." Today, there are hundreds of 5Rhythms® teachers in over 26 countries.[4] Ecstatic dances have also been influenced by the global network of electronic dance events that are held on beaches and in deserts. These are explicitly set apart from and contrasted to the cities where most participants come from, such as Boom Festival in Portugal (St. John 2009) and the Burning Man Festival in the United States (Pike 2010). The presence of ecstatic dance at Burning Man suggests one of the ways that an urban spiritual expression moves out from the city into the rural landscape in which the Burning Man event creates a temporary city in the Nevada desert. Every day for the week-long festival at "Rhythm Wave Camp," dancers can "sweat their prayers" in view of wide open spaces and barren mountains, moving and interacting in the same ways they would in downtown Oakland, referencing family and tribe and following similar guidelines to sanctify the space.

Many urban ecstatic dance events take place on Sunday mornings, as explicit alternatives to other kinds of religious services, although they are scheduled on all days of the week. Typically, there is no explicit religious or spiritual identity linked with dance events. Participants come from Buddhist, Christian, Jewish, Pagan, agnostic, and any number of other religious backgrounds, such as the catch-all "spiritual." One DJ referred to their community as "New Age, but most see no need to label the movement beyond that."[5] Although it has been significantly shaped by electronic dance music, ecstatic dance shares many characteristics, such as a focus on personal spiritual transformation, with the tradition that scholars have labeled "New Age" (Pike 2004; Heelas 1996) or more recently, "metaphysical" (Bender 2010; Albanese 2007). While dancers themselves sometimes invoke the term "New Age" to describe their identity, they are more likely to say they are "spiritual" and then describe beliefs, which usually situate them within the traditions that scholars have labeled New Age or metaphysical. DJ Medicine Heart, one of the organizers of Soul Shake Dance Church in northern California, exemplifies this New Age context. According to his biography, DJ Medicine Heart's "love for dance was born in the British DIY free party scene at the start of the '90s. He then trained internationally over the next two decades with teachers from zen, advaita, indigenous shamanic, mystical Christian, and therapeutic traditions."[6] Stacey Butcher, another northern California dance organizer, is also "a Reiki Master/Teacher, Acupressure Practitioner, Shamanic Healer and Priestess."[7]

Dancers across the United States, not just on the West Coast, share these eclectic spiritual pursuits that scholars typically locate under the New Age or metaphysical umbrella. In Dallas, Texas, for example, a website for "New Age Spirituality Meetups" lists Ecstatic Dance Dallas events alongside links for classes on meditation and "Metaphysics and Tarot," as well as a book club on "Angels and Celestial Beings."[8] According to religious studies scholar Albanese, two central characteristics of metaphysical religion are beliefs that mind and universe are dynamic and full of energy and a "yearning for salvation understood as solace, comfort, therapy, and healing" (2007: 13–15), language that many, though by no means all, ecstatic dancers would be sympathetic with. Albanese argues that an important development in twenty-first-century metaphysical religion is that the mind acquires a body, so that emphasis increasingly lies on the "enlightened body-self" (2007: 514). Nowhere is this more true than in ecstatic dance, which along with yoga and transformational festivals is one of the most significant developments in the spread of metaphysical religion in the United States in the twenty-first century.[9] All three exemplify what sociologist Courtney Bender has described as "spirituality," that is, contemporary practices that do not fit easily with traditional definitions of religion because they are "sometimes spiritual, sometimes religious, and sometimes secular."[10]

Ecstatic dances are open to anyone, regardless of their spiritual or religious identity, or lack thereof, and are characterized by what raver Jimi Fritz calls "epiphanies of acceptance," referring to communities like raves that are accepting and inclusive of gender and ethnic diversity (St. John 2009: 41). Ecstatic Dance Oakland's website promotes inclusivity: "Laugh, play, cry, skip, twirl, fly, and sashay around—knowing any expression and all movement are welcomed and encouraged in the space. All ages, genders, people of color, lesbian, gay, bisexual, transgender, queer, questioning, intersex & same-gender loving movers welcomed."[11] Most of the dance events I attended included slightly more women than men, which is typical of other New Age practices such as channeling (Brown 1999).[12] The majority of DJs are men, probably reflecting trends in rave and electronic dance music scenes. Dance events held in urban spaces like Oakland's Tropicana Ballroom reflect the diversity of the community outside; although the crowd the Sunday morning I attended was, nevertheless, disproportionately white compared to the urban Oakland neighborhoods surrounding the Tropicana.

The deeper histories of dance sites

Like other kinds of religious gatherings, ecstatic dance events are explicitly and consciously set apart from daily life. The sites dance organizers rent help to create this sense of a different kind of space away from ordinary work and home life. For instance, dances are often held in urban buildings—old businesses, ballrooms, even an old bathhouse—that figure in the cultural and financial histories of their cities. The Tropicana Ballroom, where Ecstatic Dance Oakland takes place on Sunday mornings, was once "Sweet's Ballroom," opened in 1929 by brothers William and Eugene Sweet, who loved to dance.[13] It featured famous big band acts like Count Basie, Duke Ellington, and the Benny Goodman Orchestra, as well as Latin jazz and Mexican bands playing for families on Sunday afternoons. It became a hugely popular nightlife venue throughout the big-band era and World War II when mambo kings Tito Puente and Xavier Cugat also played there.[14] During the big band era, Mexican bands would play for Sunday afternoon *tardeadas*; whole families would fill the ballroom where "native and foreign-born alike intermingled to the latest tunes from both sides of the border," according to an interview by the Latin History Project with Eduardo Carrasco, a former attendee.[15] The ballroom went through many incarnations and then, in the late 1990s, theologian Matthew Fox's "Techno Cosmic Masses," which combined sermons with dance and music, were held there. In 2010, Sweet's was the site of a shooting during a Halloween party that injured nine people and was shut down for a time.[16] Now every Sunday morning, Ecstatic Dance Oakland hosts dance events for participants from across the Bay Area. Most ecstatic Oakland dancers make no reference to these other histories nor to the high-crime neighborhoods nearby. They filter out some aspects of their inner-city location—in order to maintain a sense of the dance space as a sanctuary—while welcoming in others. Yet traces of the ballroom's Jazz Age past and colorful history linger and heighten the sense of being in a city with a rich musical heritage, connected to but set apart by its red carpets and faded elegance from violent episodes of the past when urban Oakland intruded the space.

Across the San Francisco Bay in San Francisco's Mission District, Ecstatic Mission, held in a basketball gym at Mission Dolores Academy, harbors histories of another, older era. Next door to the gym is the oldest mission church in California and the oldest building in San Francisco, where hidden behind the main altar is a mural painted by Ohlone and other "Mission Indians" in 1791.[17] This visual trace is a reminder of the indigenous people who once lived and danced on the lands where Mission Dolores stands today; it recalls the impact of European religion and colonialism on these indigenous communities. The mural is also a reminder that world renewal dances were once held throughout this part of California, the kind of "primitive" past ecstatic dancers invoke as a model for their own spiritual practice. At dance events, there are often generic references to so-called "tribal" cultures but not to the specific people who lived in the lands where dances are held: the Miwok, the Ohlone, the Patwin (Margolin 1978). By referencing generic "tribal people," San Francisco dancers don't have to acknowledge the impact of the Mission on the Ohlone and other Native Californians. By selective remembering, they can avoid opening up the space to a painful and complicated history that works against simplistic ideals of the generic tribal past that they invoke for inspiration when they dance.

Across the United States, other ecstatic dance events also occupy buildings in older urban neighborhoods with deep histories. Six hundred miles north, Sacred Circle Dance, the largest ecstatic dance gathering in Portland, Oregon, takes place every Sunday morning at the Tiffany Center, an Art Deco building, built in 1928, and listed on the National Register of Historic Places that features an entranceway with a local artist's reproduction of Michelangelo's Sistine Chapel and stone-carved gargoyles.[18] In 2011, Ecstatic Dance Providence was held at Providence, Rhode Island's Butcher Block Mill, once the United States Rubber Company Mill (1896–1910), a golf ball factory and a furniture mill before being renovated into an artist community and small business space. In Dallas, Texas, Dallas Ecstatic Dance held a New Moon "Ecstatic MoonDance" in May 2014 at the Dallas Bath House Cultural Center. Built in 1930 on what were then the rural shores of White Rock Lake, the "old bath house" was one of the first uses of Art Deco architecture in the Southwest. Made over into a cultural center, it reopened in 1981 and now houses a theater, art galleries, and the White Rock Lake Museum.[19] These urban sites play a role in dancers' imaginings of the possibilities for transformation that they believe dance events offer. Like the middle-class nineteenth-century Americans that Robert A. Orsi discusses in his introduction to *Gods of the City*, dancers approach these urban sites "as places where they might explore dimensions of their identities . . . that were otherwise and elsewhere religiously and socially constrained" (Orsi 1999: 9). Dancers' expectations about what they will experience at ecstatic dance gatherings are heightened by traveling away from home to sites with subtle echoes of distant eras and histories.

While secular spaces like these old ballrooms, gymnasiums, industrial buildings and cultural centers are the most common sites for ecstatic dances, in New York City, regular dances are held at St. Mark's Church in-the-Bowery, a progressive Episcopal church in the East Village neighborhood of Manhattan. St. Mark's, the oldest site of continuous worship in New York City and the burial site of Peter Stuyvesant and other founding families of New York, was completed in 1799 and, like some other dance spaces, has had a colorful history, "as long and meandering as the history of European settlement in Manhattan."[20] As one dancer describes it, "St. Mark's . . . [is] a beautiful venue to have the event, symbolic, historic, open, prayerful . . . it's really quite perfect."[21] Dancers at St. Mark's, as at other historic sites, may be vaguely aware of the histories that shape their experience of dance space as special and they invoke the "historic" charm of these spaces. Oakland dancer Grier Cooper explains why Oakland Ecstatic Dance at Sweet's Ballroom is one of her favorite places to dance in the San Francisco Bay Area:

> I have a thing for historic buildings, particularly those with architectural details that showcase true craftsmanship. Whenever I'm in an old building I like to imagine what it was like long ago. It's as if the walls still whisper echoes of secrets. Back in the Big Band era of WWII, Sweet's was the go-to venue in Downtown Oakland.[22]

At St. Mark's, dancers tread the floors where many famous artists, writers, musicians and dancers came before them. In the past century, St. Mark's supported and hosted poet William Carlos Williams, singer/songwriter Patti Smith, Houdini, and even the famous dancers Isadora Duncan, who danced in the church in 1922, and Martha Graham in 1930.[23] These urban dance sites carry hints of deeper histories of manufacturing and labor disputes, colonialism and gentrification, crowded tenements and Jazz Age decadence, not acknowledged by dancers but, nevertheless, part of an aura that makes these special spaces.

How does it matter that Benny Goodman performed at Sweet's or Martha Graham at St. Mark's? Dance communities make few references to the rich pasts of the spaces they dance in even when there is a dance connection, as in the case of Isadora Duncan and Martha Graham at St. Mark's. But dancers and the historic spaces where they dance might be seen in a kind of reciprocal relationship. These buildings link them to the past and provide an aura of authenticity, or as Grier Cooper puts it, "the walls still whisper echoes of secrets," but dancers also bring new life into old buildings and their neighborhoods. In the case of Ecstatic Dance Oakland, dancers bring music and a festive atmosphere to Sweet's Ballroom, as well as funding to help maintain the building.

The pasts that are most explicitly invoked at ecstatic dance events involve an imagined tribal past, in which dance and ritual was central to community life and the past of the individual self that they believe needs to be transformed through movement. In order to "forget" a shooting at Sweet's a few years back or the history of colonization and Christianization at Mission Dolores, to make over secular urban spaces into sanctuaries for spiritual experience separate from the city outside and the deeper histories of the neighborhoods where dances are held, dancers employ a number of different strategies of purification and sacralization involving sound, colorful lighting, material culture and regimes of the body.

Creating safe and sacred space

To enter the dance sanctuary, participants must pass by a table where someone sits to take payments and prevent homeless and other non-paying individuals from entering. This boundary precludes just anyone from walking in off the streets and alerts dancers to the fact that they have entered a carefully bounded space. Ecstatic dance events are usually open to the public and publicized through Facebook pages and websites, as well as word of mouth.[24] Publicity stresses the sanctuary aspect of the dance space and rules about no alcohol, no drugs, no talking. Belonging to the dance "tribe" is as simple as showing up and following a few guidelines: tribal belonging is defined by the physical space and organizers' intentions rather than any initiatory or vetting process. When Ecstatic Dance Providence organizers released an announcement for their Winter Solstice event, they exhorted dancers to "leave your holiday stress and drama at the door."[25] Dancers are told to decide what to bring with them and what to leave outside the dance space. By decorating unexpected worship spaces, ecstatic dancers signal that these gyms and ballrooms are serving a different purpose: this is not a high school basketball game, nor a salsa dance night. They make over these sites in a number of ways to let participants know they have entered a space outside of history and away from their daily work and family lives, as well as the bustle of the city. Ecstatic Mission dance organizers hide Mission Dolores Academy gymnasium's basketball hoops behind bright banners and place strings of LED Christmas lights in a colorful perimeter along the edges of the basketball floor; most dance organizers have strict rules that cleanse dance spaces of smoke, alcohol, scents, talking and trash; many dance events use lights, banners and sound, and create altars for silent or whispered informal prayer. All the strategies that dancers employ to temporarily transform these urban spaces help them create a sense of a bounded community that they feel is safer than the city outside and, in which they expect transformational or spiritual experiences to happen.

Just as participants approach a visible entryway where greeters (and bouncers in some cases) wait to usher them inside, so too must they cross an aural boundary as street sounds give way to electronic music that invites them to enter and dance. Music is an important tool for dancers' work of making over ordinary spaces into sanctuaries of sound for free-form worship. DJs are seen as shamans of sound, tasked with taking the room of dancers to a state of prayerful ecstasy. As Ecstatic Dance Dallas organizers promise in online promotion of their event:

> These sonic shamans blend live ethnic percussion remixes of the hottest tracks from the best tribal dub and world electronic dance music artists and bring it to life with visionary projected art . . . taking the listener/dancer on an exotic, ecstatic journey.[26]

Although dance events usually have no leaders, ministers, gurus, or priestesses, DJs with names like "DJ Sweat Miracles," "Medicine Heart," "Baron von Spirit" and "DJ Yogi" are responsible for dancers' experiences, playing music designed to lead them through a kind of ritual process that builds to an ecstatic climax. Some DJs are solely focused on music, while others, like dancers themselves, may participate in related spiritual practices. DJ Sharu, from the San Francisco Bay Area, who was a special guest in 2012 at Ecstatic Dance NYC held in St. Mark's on the Bowery, calls himself a "Vedic astrologer and Yogi." Ecstatic Dance NYC organizers promised participants that DJ Sharu would infuse "his sets of next-level, movement-inspiring electronic dance music with just enough sacred to rock our worlds."[27] DJs, then, must balance "just enough sacred" against any sense that they are proselytizing. They are expected to facilitate dancers' experiences of having their worlds "rocked," but not act as spiritual leaders.

Sound at dance events includes not only music, but also chanting, whistling, shouting and laughing. However, talking is not allowed: organizers of the Los Angeles dance gathering "Fumbling Toward Ecstasy" remind participants:

> You are entering Sacred Space as you come into this room, and your observance of this is important to everyone present. If you feel called to talk with someone please take any verbal conversation out of the room, or convert it to movement.[28]

Most dance events designate other spaces for talking outside the room where the dance is held. More typically, people save their talking for after the dance has ended, when food and drink, as well as conversation, are sometimes shared together. As Isaac Weiner points out in his 2013 study *Religion Out Loud*, "*how* religions have sought to make themselves heard has always been at least as important as *what* they were trying to say" (2013: 200, original emphasis). At ecstatic dance gatherings, movement, music and dancers' vocal expressions (but no talking) generate dancers' experiences. According to Dwain, a Dallas dancer:

> As a drummer, the primal beat has always found my soul and pulled my body into motion. If it pushes out all sides of my spirit, then I euphorically flow into a semi-oblivious trance . . . and depending on the music and rhythm either my beast or ballad comes out on the dance floor.[29]

Dancers find that they make different sounds and move differently at ecstatic dance events than in their lives in the outside world. Dance allows them to cleanse themselves of the city's detritus, or whatever in their lives they think is not serving them. A participant at "High Vibration Waves," a dance event in Manhattan, noted that "Dancers occasionally released guttural howls, as if exorcising the demons of the workweek."[30] The music and other sounds of ecstatic dance separate dance spaces from other urban soundscapes. Sound and music offer dancers relationships they can participate in and use to generate cathartic and healing experiences on the dance floor that they say help them better deal with their work and family back home.

Temporary altars are another way that dance organizers signal to dancers that they have entered a special space, separate from the urban spaces outside and yet, at the same time, connected to larger religious traditions. There is typically no structured worship at ecstatic dance events and no one thing or being all dancers would point to as their spiritual inspiration; yet altars created for each dance provide a visible focus for dancers' prayers. San Francisco's Ecstatic Mission dance organizers prepare an altar every week as a "safe space to sit and reflect, to meditate, or pray as you wish."[31] Altars may include flowers, plants, candles, tarot or medicine cards, images of Hindu deities, goddess figures or the Buddha, a picture of someone who has died or is going through a healing crisis, and sometimes fruit to share. Ecstatic dance altars also sometimes vary to reflect the seasons or are tailored to the theme of a particular dance. At an ecstatic dance gathering in Providence, Rhode Island on Winter Solstice, dancers were invited to bring objects for the altar "to energize it and get . . . energized by the ecstatic vibes." These smaller sites of focused prayer within the larger space draw on familiar religious idioms and help to further obscure the ballroom or gymnasium and heighten the contrast dancers want to make between these spiritualized spaces and secular life outside.

Because dance events are eclectic and inclusive, images of the Buddha or Shiva on ecstatic dance altars generally seem to stand in for a generic "deity" or "spirit." Dancers meditating at the altar may be praying to Buddha or God, but are, just as likely, *praying to no one in particular* for an ill friend to be healed, for the safety of a son joining the military, or for guidance on their own path ahead. Dancers describe prayer as an expression of the inner self or "divine self" through movement: dance itself is their prayer. Ecstatic Dance San Francisco organizers set the scene in this way: "You may find yourself Dancing in your Bliss . . . or Dancing your Grief . . . Perhaps with your Joy, your Sorrow . . . Your Anger, Your Silliness . . . Your Sexiness . . . All of it . . . This is Your Dance Space."[32] Spiritual experience in the ecstatic dance context often moves from the inside out, or as one New York-based 5Rhythms® dancer puts it, "This is an opportunity to investigate the unexplored sacred wilderness of bone and breath, hands and hips, spirit and flesh."[33] Dancing, like meditating before a dance event's altar, is an outward expression of an inner devotional process where the process itself is what is important, not the object at which the dancer's gaze is directed. As Gabrielle Roth explains in her book *Sweat Your Prayers: The Five Rhythms of the Soul*: "to sweat is to pray, to make an offering of your innermost self. Sweat is holy water . . . The more you sweat, the more you pray. The more you pray, the closer you come to ecstasy" (Roth 1997: 1). Dance may be an offering of self, but it is also a devotional prayer *to* self, to a self that is transcendent, embodied, and made more easily accessible because of the efforts of dancers to sanctify these urban dance spaces.

No shoes, no scents, no talking: discourses of purity

Like raves and neotrance events, ecstatic dance events have no structure and no steps to learn, but they do have rules that act to set them apart from these other occasions: no alcohol, no drugs, no street shoes, no strong scents and no talking on the dance floor. DJ Sweat Miracles explained to newcomers in a morning circle at Soul Shake Dance Church, held every Sunday in a college town in northern California, that there is no talking because the dance floor is "a place for the body and the heart to have their own language."[34] Many dances even request that dancers avoid wearing perfume out of respect for those participants who might be allergic: scents and substances of the outer world must be left outside. Ecstatic Dance Palo Alto advises dancers: "Dress to sweat, bring water, no perfumes/colognes," and Ecstatic Dance Oakland asks dancers to follow these guidelines: "no pictures, video, or wearing of scent (please be clean)."[35] Ecstatic Mission organizers also explicitly make the point that the dance space should be trash-free, asking that participants: "bring your own [water] bottles, as this is a no-waste event."[36]

"No-waste," then, means not laying waste to the body either: "This is a Sacred Space where you are invited to come as you are, however that is; without the use of alcohol or recreational drugs."[37] By purifying the body, ecstatic dancers suggest that purifying the psyche is more attainable. Dancers repeatedly draw boundaries between dancing in bars or clubs and dancing at dance church. Along with rules about no alcohol, drugs, scents, talking, or trash, most dances have explicit guidelines about physical contact. At the beginning of every Soul Shake Dance Church event, organizers gather participants in a circle and remind them that there is no talking on the dance floor and that they should feel free to dance with others, but watch carefully to see how the other person responds. The organizers let participants know that some people will be doing contact improvisation or dancing together in ways that might seem sensual or erotic; but this kind of dancing is purely a matter of choice: if they do not want to dance with someone who approaches them, they can simply put their palms together in a prayer position, meaning they would prefer to dance alone. In practice, many people dance alone most of the time or fleetingly engage with another person, meet their eyes, move with them for a few minutes, then move on.[38] Performance fueled by artificial highs and ego is to be avoided, as are dancing for other people rather than for oneself or getting into someone else's personal space in inappropriate ways. In the words of dancer Jewel Mathieson:

> Not the pretty, pick me dance. But the claw our way back into the belly of the sacred, sensual animal dance . . . Not the jiffy booby, shake your booty for him dance . . . We have come to be danced . . . In the cathedral of flesh. . . .[39]

Discourses of purity allow ecstatic dancers to distinguish their "cathedrals of the flesh" from other types of urban electronic dance where drugs are usually present, from city neighborhoods outside where women especially do not feel safe, and from drunken hook-up scenes in bars. As "cathedrals," their dancing bodies serve as both vessels of purity and symbols of spiritual possibility. Since dance events include explicit guidelines that contrast dance spaces to bar pick-up scenes, dancers learn to be sensitive to others' personal boundaries and to say "no" if anyone transgresses them. Ideally, at least, after the dance has ended, no one feels violated in any way; no one is staggering home or passed out in a corner and there are no plastic cups or glass bottles left to clean up.

Like the prohibition of polluting substances on and in the body, dancers also leave the dirt of the streets outside. "Come and dance your Dreams Awake in our playfully sacred portal at the St. Mark's Church as we invite you to peel off the layers of the streets to reveal and celebrate your passion and divine eternal Self!"[40] Street shoes are not allowed on the dance floor, and though dance shoes and socks are usually permitted, bare feet are more common. Taking off street shoes is both a literal and symbolic removal of the outer world's dirt at the same time that it suggests another kind of relationship between feet and floor. As one dance organizer explains: "Being Barefoot is also encouraged to feel more connected to the ground," even though the ground is many layers down under the building's foundations.[41]

Dancers repeatedly describe the secular spaces where dances are held as "safe" and "sacred" sanctuaries apart from the city, its crime and nightlife: for Cynthia, a Seattle dancer, "It's a wonderful safe space to dance through the pressures of the world and emerge into loving ecstatic expansiveness."[42] At ecstatic dance events, it is a sacred self that is the goal. It is not possession but the kind of "dispossession" that anthropologist Graham St. John argues characterizes neotrance dance festivals: "a dispossession. . . . where trancers find relief from a troubling and dispirited lifeworld" (St. John 2009: 50). Ecstatic dance purports to be an escape from the sights, sounds, smells and polluting properties of the city and selves outside dance spaces, but as hard as dancers try to escape, in many ways dance events remain very much rooted in the urban landscape.

Porous boundaries: what dancers bring with them and take back into the city

Ecstatic dance events exist within complex networks outside the boundaries of the temporary dance sanctuaries that organizers and participants create. As Orsi points out in *Gods of the City*, "No aspect of urban experience is impermeable to others" (Orsi 1999: 57). For instance, dancers participate in both local and global communities as well as in their dance congregations. Many dances set aside a time at the end of the dance for "community announcements," and include tables with promotional materials for other local events: flyers about spiritual retreats, life-coaching, improvisation classes, yoga classes, psychotherapy and massage therapy that are happening around the cities in which dances take place. Like the "neo-nomad" techno and New Age movements described as "global circuits of countercultural lifestyle," in sociologist Anthony D'Andreas's *Global Nomads: Techno and New Age as Transnational Subcultures* (2007), dancers partake in larger networks that form what has become for them a "tribe" of like-minded others around the world. Worldwide synchronized dance events like Earthdance and huge eclectic festivals like Burning Man in the Nevada desert are explicitly mentioned on dance sites and information about them shared among participants, although these are sites at which alcohol, drugs, or dirt are allowed. Urban dance events with their sanctified spaces, then, are both separate from (because of their purity practices) and connected to these other dance sites through participants who circulate through these other event spaces as well as share stories about them. Dancers thus move through interconnected networks that include the safe and set-apart spaces of dance events and, at the same time, link particular dance floors to other practices and gatherings that focus on spiritual transformation and express characteristics of New Age or metaphysical religion, as described by scholars like Catherine Albanese and Courtney Bender (Albanese 2007; Bender 2010).

Describing the dance community as tribe or family is another way that dancers heighten their expectations about what they will experience at dance gatherings and distance themselves from city streets at the same time that they connect to a larger network of people and communities outside dance space. 5Rhythm's official website describes "our tribe" in this way: "We are beat-driven, service-oriented, heart-based individuals who come together to embrace our tribal longings."[43] Ecstatic dancers emphasize their ancient roots in shamanistic traditions that they believe captured a blend of transcendent ecstasy and embodied prayer. Karen Berggren sees ecstatic dance as the return of earlier spiritual expressions. She argues that dance has been cut off from its "healing and esoteric" roots and made into "a way to socialize, or to attract a mate," rather than a way to pray. But in the past decade, she notes, "it seems that the ancient understandings of dance as a healing and spiritual modality are impressing themselves upon the modern mind, as if welling up from old, vibrant memories of the tribal soul within."[44]

At contemporary ecstatic dance events, dancers like Berggren search for what anthropologist Graham St. John calls "sacred sociality" in the context of electronic dance music cultures (St. John 2009: 38). In much writing about the New Age movement, emphasis is put on techniques of the self, what sociologist Paul Heelas calls the spirituality of the self, but discourses of the self in New Age communities are often in tension with the emphasis on tribe, family and community (Heelas 1996). This tension is particularly notable at urban dance events when city life is set in contrast to nostalgia for a supposedly simpler and healthier tribal lifestyle. Many dancers live in the tension between their professed ideal of a simpler, more fulfilling "tribal" life and the reality of their urban existence. Discovering like-minded others with similar ideals within the urban context, then, gives them a sense that they are creating alternative communities, a tribe, even while going about their lives in their urban neighborhoods and work environments. In her testimonial on the 5Rhythms® website, dancer Helena Kallner explains that through ecstatic dance, "I have also found a community, a worldwide family of dancers."[45] Dance events, then, become spaces in which dancers experience being temporarily connected to their ideal community, which they understand as "home" or "tribe."

At the same time that they embrace modern technology in the form of elaborate sound systems, mix tapes and social media, dancers have inherited the Romantic tradition of looking to the past as a model for all that is missing in today's world. They often speak of the past with a sense of loss and nostalgia, invoking ancestral cultures that they want to refashion for contemporary people. As dancer Joey sees it, "The desire to dance feels like an ancient current in my body, a deep knowing that this is my way to connect with the divine."[46] In a talk she gave at the Burning Man Festival, dance teacher Shiva Rea described her vision of a sacred history when humans everywhere danced around fire—"things we've forgotten"—followed by the repression of dance by Christianity and the recent revival of ecstatic dance.[47] Kathryn, a participant in Ecstatic Dance Seattle, thinks of her dance community as a family that welcomes her "home" every time she dances, so that she feels "a part of something much larger than myself." When dancers "come home," they understand home as a spiritual reality inscribed in physical spaces, channeled through the body, and discovered in one's own consciousness within a community of like-minded others:

> I feel a part of something much larger than myself . . . Born of a mixture of light-skinned European ethnicity. I've always wanted to belong to a culture, a tradition, a tribe . . . I was greeted by another dancer with . . . a whisper in my ear, "Welcome home."[48]

In Kathryn's case, ecstatic dance is more than movement or self-expression; it connects her to other people and cultures in ways her own background does not.

Dancers not only reference an imagined tribal past, but also their personal pasts, invoking childhood expression and experience on the dance floor. Dancers bring their pasts to the dance floor, especially the playfulness of childhood, and, at the same time, they attempt to transform family traumas and heal childhood wounds. They both cut themselves off from the past and restore the past to themselves, remade by dance, an ideal past they have created. Dancers bring their inner histories—childhood memories and family backgrounds—with them to dance events, creating new expressions of spirituality from inherited and borrowed traditions, including using Christian language of church and prayer from a tradition they have often rejected. Dancers are Buddhist, Christian, Jewish, Pagan, agnostic, or "spiritual," and many of them fall into the "None" category on recent American religious identification surveys.[49] But they transpose the religious experiences they grew up with and may have practiced elsewhere onto ecstatic dance sites. They bring with them memories of religious rituals they visited or participated in when they were children, images of other people's gods, blessings and prayers from their families or other nonreligious family traditions. Dwain identifies his family's love of dancing as the heritage he brings to dance church:

> My late grandmother competed in national ballroom competitions and taught me many varieties of dance . . . During my adolescent years all of us would always go two-stepping, waltzing, polkas . . . My parents were regular members of a Dallas dance club So dancing has always been in mine and my family's DNA.[50]

Many dancers, then, look to the past, their family traditions and childhood experiences for inspiration, bringing these other worlds of memory and experience with them to the dance floor.

If the inner histories of dancing bodies are tools of continuity between the world outside and the sacred dance floor, then bodies too connect dancers and urban dance spaces to nature, reminding dancers that even in the city where the soil of the earth is many layers below the dance floor, they can still connect through movement and imagination to the natural world. Meaghan, a New York City dancer, explains that 5Rhythms® dancing allowed her to

> . . . engage with the earth . . . After many years of seducing my edge on the city ledge, I leapt, landed, and followed my feet down and way out to the actual-star-studded critter-crawling Earth, where I've found an essential touchstone of my movement practice in nature. For me it's profoundly healing and humbling to dance consciously in and with the natural environment.[51]

On "Why Dance," a space for testimonials on the Oakland Ecstatic Dance website, Oakland dancer Onyx explains that for her, dancing links human bodies to the Earth's body:

> When i dance i'm in my body, which is where i want to spend more time . . . I also can sometimes feel Gaia, our extended earth body—our connection to the entire web of Life—through experiencing the elements in the bodily senses.[52]

Many dance communities honor the cycle of the seasons with special Equinox or Solstice themes. By connecting their own movement to the cycle of seasons, dancers emphasize the ways in which their bodies extend into the bodies of others, into earth, animals and peoples of the past that they invoke through the language of tribal belonging. Love RahKen, a San Francisco dancer, sees it this way: "It's my art, my prayer, my sanity. Being human can be a masquerade of sorts . . . dance reminds me that I am an animal . . . a pack, a pride, a tribe."[53] When dancers transpose the language of tribal life in nature onto urban landscapes, they simultaneously separate dance from other spaces *and* link dancing to other bodies (earth and animal) and other forms of movement and spiritual expression.

Transforming densely layered urban spaces into temporary temples for dance requires selective remembering of particular pasts and not others. Dancers call into question boundaries between human and nonhuman animals, individual self and communal belonging, child and adult, city and nature. Organizers of a London "ecstatic trance dance" reminded participants, "We all share DNA with all the animals alive on earth. We can in our memories see out of any animal's eyes. We are them."[54] Ecstatic dance events reflect these and other relationships that are often characterized by tensions and contradictions. Dancers disregard the deep histories of the sites where they dance, but invoke a primordial tribal past. They are drawn to the anonymity of the city and yet they focus on dance events as tribe, family and home, craving a sense of an intimate community with other fellow spiritual seekers. Their roots are clearly urban in that the dance techniques they draw from, like Gabrielle Roth's 5Rhythms®, emerged out of and were nurtured by urban communities, even as they spread to smaller college towns and suburbs. Dancers invoke nature and primitive culture and yet are immersed in fast-paced city life and the many aspects of technology that they use to network with other dances and events. They romanticize tribal communities and yet say nothing about the actual indigenous people who populated the sites where they now dance. They work hard to separate dance spaces from the city, but they bring other experiences and communities in with them and take their dance experiences out onto the streets, spiritualizing the city as they go. They step over the boundaries they have so carefully created and take their dancing bodies into neighborhoods and public parks, interacting with people and communities they meet there. After Ecstatic Oakland ends every Sunday morning, dancers put away their hula hoops, pick up their water bottles, take down the altar they created, and put on their street shoes. They make their way down the ballroom stairs, pick up flyers for yoga retreats and other dance events on their way out, and head to a dance community picnic at downtown Oakland's Adams Park to "make friends, play, sing songs, eat food, enjoy the outdoors," partaking in the life of the city around them and bringing relationships and experiences forged in dance sanctuaries with them into other urban spaces.

Notes

1 Although ecstatic dance is a global movement, this essay focuses on the US, where I have conducted fieldwork.
2 "Tending the Sacred Fire: The Origins, Repression and Revival of Free Form Movement," Burning Man Festival's "Ted X Black Rock City," https://www.youtube.com/watch?v=i31gGh9KcAo; accessed December 12, 2014.
3 http://ecstaticdance.org/origins/; accessed April 1, 2015.
4 Jed Lipinski, "Dance, Dance, Dance. And That's It." *New York Times*. August 5, 2010. Http://www.nytimes.com/2010/08/05/fashion/05Sober.html?_r=0; accessed March 7, 2014.
5 Field notes, February 9, 2014.
6 http://transformationalinc.net/soul-shake-ecstatic-dance-wave-with-luke/; accessed April 5, 2015.
7 San Geronimo "Sweat Your Prayers," http://sweatyourprayerssg.com/the-5rhythms-practice/; accessed April 5, 2015.
8 http://new-age-spirituality.meetup.com/cities/us/tx/dallas/; accessed April 5, 2015.
9 Elizabeth Perry, "Transformational Festivals: where Ecstatic Spirit and Sonic Celebration Unite," http://www.redefinemag.com/2013/transformational-festivals-spiritual-preview-guide/. Dancers across the country often comment on the movement's growth. See the "Ecstatic Dance Community" website: http://www.ecstaticdance.org; accessed April 17, 2015. I witnessed the evolution of Soul Shake Dance Church in Chico, CA, my hometown. In its first year, twenty dancers at most might show up each Sunday. Eight years later in 2014, it attracted eighty to a hundred dancers every week.
10 "Spirituality Entangled: An Interview with Courtney Bender," http://blogs.ssrc.org/tif/2010/06/01/spirituality-entangled/; accessed August 15, 2016.
11 http://www.awakening360.com/business/ecstatic-dance-oakland#sthash.c5k1Iii9.dpbs; accessed March 25, 2014.

12 More quantitative research would need to be done to see if this is typical across the country. The dances I went to in smaller communities had fewer men in proportion to women, while the dances in Oakland and San Francisco, with their large gay, lesbian, bisexual, transgender and queer (GLBTQ) communities, included more men. Because 5Rhythms® dance and other dance techniques often encourage participants to explore both "feminine" and "masculine" sensibilities, defining participants with this gender binary may be misleading. At the Oakland and San Francisco events in particular, participants may self-define as queer or transgender.

13 Kimberly Chun, "Sweet's Ballroom Swings Again As Part of Oakland's Renaissance," *San Francisco Chronicle*, April 30, 1999, http://www.sfgate.com/bayarea/article/Sweet-s-Ballroom-Swings-Again-As-Part-of-2933643.php; accessed March 25, 2014.

14 Angela Woodall, "Night Owl: La Tropicana in Oakland replaces historic Sweet's Ballroom and brings back Latino flair to downtown," April 28, 2013, http://www.insidebayarea.com/oakland-tribune/ci_23117261/night-owl-la-tropicana-oakland-replaces-historic-sweets; accessed March 12, 2014.

15 Ibid.

16 http://www.insidebayarea.com/news/ci_18233545; accessed March 13, 2014.

17 "Centuries-old murals revealed in Mission Dolores/Indians' hidden paintings open window into S.F.'s sacred past," *San Francisco Chronicle*, January 29, 2004, http://www.sfgate.com/news/article/Centuries-old-murals-revealed-in-Mission-Dolores-2804668.php; accessed December, 10, 2014.

18 http://www.tiffanycenter.net/?page_id=2; accessed December 12, 2014.

19 http://www.dallasculture.org/bathHouseCultureCenter/aboutHistory.asp; accessed April 2, 2014.

20 http://stmarksbowery.org/welcome/about/; accessed April 2, 2014.

21 http://sensualtantrichealing.wordpress.com/2012/09/20/ecstatic-dance-nyc-1/; accessed April 2, 2014.

22 "5 Best Places to Dance," http://www.griercooper.com/?tag=sweets-ballroom; accessed April 10, 2015.

23 http://stmarksbowery.org/welcome/about/; accessed April 2, 2014.

24 Sometimes they are called "drop-in" dances to distinguish them from more formal classes in 5Rhythms® that usually require a significant time commitment, such as twelve weeks. These classes usually involve more active recruitment since the teachers are reliant on them for income. Regular drop-in dance events are not seen as money-making opportunities, although organizers hope to recover costs for renting the space. However, dancers may first learn about the more structured dance classes from flyers and announcements at drop-in dances. For example, Jacia Kornwise runs a dance studio in which she offers a drop-in Tuesday night dance open to anyone who wants to show up, as well as structured classes and retreats only open to those who have registered in advance (https://www.facebook.com/SatoriCenter/events; accessed August 15, 2016). She leaves her flyers at the entrance to Soul Shake Dance Church, where there is a Sunday morning drop-in dance.

25 Ecstatic Dance Providence Facebook page, https://www.facebook.com/EcstaticDanceProvidence; accessed March 20, 2014.

26 http://www.meetup.com/Ecstatic-Saturdays-Dallas/; accessed April 1, 2014.

27 Ecstatic Dance NYC event listing, http://www.eventbrite.com/e/ecstatic-dance-nyc-st-marks-church-tickets-3786942848; accessed April 9, 2015.

28 http://www.movinground.com/#!about/cuyo; accessed March 25, 2014

29 http://www.meetup.com/ecstatic-dance-dallas-movement-music-community/messages/boards/thread/29691622; accessed April 2, 2014.

30 Jed Lipinski, "Dance, Dance, Dance. And That's It." *New York Times*. August 5, 2010. http://www.nytimes.com/2010/08/05/fashion/05Sober.html?_r=0; accessed March 7, 2014.

31 http://alistcalendar.com/2013/03/ecstatic-dance-sf/; accessed March 7, 2014.

32 http://www.indiegogo.com/projects/ecstatic-dance-san-francisco; accessed March 7, 2014.
33 5Rhythms®, http://www.5rhythms.com/classes/UrbanWavesBytheBook-788; accessed August 15, 2016.
34 Field notes: Soul Shake Dance Church, Chico, CA, 23 February 2014.
35 http://www.consciousdancer.com/events/ecstatic-dance-palo-alto/and http://ecstatic dance.org/guidelines/; accessed April 10, 2015.
36 http://www.indiegogo.com/projects/ecstatic-dance-san-francisco; accessed April 10, 2015.
37 Ecstatic Dance San Francisco, https://www.indiegogo.com/projects/ecstatic-dance-san-francisco; accessed April 10, 2015.
38 Field notes: Soul Shake Dance Church, Chico, CA, 23 February 2014.
39 http://www.meetup.com/ecstatic-dance-dallas-movement-music-community/; accessed April 2, 2014.
40 http://www.sharu.us/2012/07/13/sharu-ecstatic-dance-nyc/; accessed 4 April 2014.
41 http://www.indiegogo.com/projects/ecstatic-dance-san-francisco; accessed March 25, 2014.
42 http://www.ecstaticdanceseattle.com/testimonials/; accessed March 26, 2014.
43 http://www.5rhythms.com/who-we-are/5rhythms-onetribe/; accessed March 25, 2014.
44 Karen Berggren, "Ecstatic Dance FAQ," http://www.firetribehawaii.org/articles/ecstat icdance.shtml; accessed March 20, 2014.
45 http://www.5rhythms.com/who-we-are/5rhythms-onetribe/; accessed December 12, 2014.
46 http://ecstaticdance.org/whydance/; accessed March 15, 2014.
47 "Tending the Sacred Fire: The Origins, Repression and Revival of Free Form Movement," Burning Man festival's "Ted X Black Rock City," https://www.youtube.com/watch?v=i31gGh9KcAo; accessed December 15, 2014.
48 http://www.ecstaticdanceseattle.com/testimonials/; accessed March 25, 2014.
49 Pew Research Religion and Public Life Project, "'Nones' on the Rise," October 9, 2012, http://www.pewforum.org/2012/10/09/nones-on-the-rise/; accessed December 14, 2014.
50 http://www.meetup.com/ecstatic-dance-dallas-movement-music-community/messages/boards/thread/29691622; accessed April 2, 2014.
51 http://www.5rhythms.com/; accessed March 21, 2014.
52 http://ecstaticdance.org/whydance/; accessed March 20, 2014
53 http://ecstaticdance.org/whydance/; accessed March 25, 2014.
54 http://www.meetup.com/londonspiritualcentre/events/64602402/?action=detail&eventId=64602402; accessed April 17, 2015.

References

Albanese, Catherine L. (2007), *A Republic of Mind and Spirit: A Cultural History of American Metaphysical Religion.* New Haven, CT: Yale University Press.

Bender, Courtney (2010), *The New Metaphysicals: Spirituality and the American Religious Imagination.* Chicago, IL: University of Chicago Press.

Brown, Michael F. (1999), *The Channeling Zone: American Spirituality in an Anxious Age.* Cambridge, MA: Harvard University Press.

D'Andreas, Anthony (2007), *Global Nomads: Techno and New Age as Transnational Subcultures.* New York: Routledge.

Heelas, Paul (1996), *The New Age Movement: The Celebration of the Self and the Sacralization of Modernity.* Oxford: Blackwell.

Margolin, Malcolm (1978), *The Ohlone Way: Indian Life in the San Francisco-Monterey Bay Area.* Berkeley, CA: Heyday Books.

Orsi, Robert A. (1999), Introduction. Crossing the City Line, in Robert A. Orsi (ed.), *Gods of the City. Religion and the American Urban Landscape*. Bloomington: Indiana University Press.

Pike, Sarah M. (2004), *New Age and Neopagan Religions in America.* New York: Columbia University Press.

Pike, Sarah M. (2010), Desert Goddesses and Apocalyptic Art: Making Sacred Space at the Burning Man Festival, in Katherine McCarthy and Eric Mazur (eds), *God in the Details: American Religion in Popular Culture.* New York: Routledge.

Roth. Gabrielle (1997), *Sweat Your Prayers: Movement As Spiritual Practice, the Five Rhythms of the* Soul. New York: Jeremy P. Tarcher.

St. John, Graham (2009), Neotrance and the Psychedelic Festival, *Dancecult: Journal of Electronic Dance Music Culture*, 1: 35–64.

Weiner, Isaac (2013), *Religion Out Loud: Religious Sound, Public Space and American Pluralism.* New York: New York University Press.

10 A saxophone divine

Experiencing the transformative power of Saint John Coltrane's jazz music in San Francisco's Fillmore District

Peter Jan Margry and Daniel Wojcik

Although raised in a white Protestant milieu, Oriah Vaughn (age 44) from Virginia no longer attends church services or identifies as a Christian, explaining that she has "cobbled together my own spiritual beliefs from a variety of religions . . . I believe every religion has something to teach us." Like many other urbanites, her spiritual pursuits connect to self-realization and reflect the worldwide process of the subjectivation of present day "believing" (Heelas and Woodhead 2005). As a jazz enthusiast, Oriah visited the Saint John Coltrane Church in 2011 during a trip to San Francisco and described her immersion in the soundscape of Coltrane's music as

> . . . an expression of worship that was moving beyond any other experience I've ever had during any other church experience. Everyone was filled with joy and expressed that to everyone who walked in the doors. I felt spiritually lifted and elated . . . I left the sermon in a slightly altered state of consciousness.[1]

Oriah is not alone in her spiritual experience of Coltrane's music, as many other devotees of jazz are familiar with the religious and transcendent feelings evoked by Coltrane's numinous "sheets of sound." The relocation of the Saint John Coltrane Church into a hip and increasingly gentrified part of the Fillmore district now accommodates increasing numbers of tourists and jazz enthusiasts in a safe and inviting environment. The Coltrane celebrated in this context is not only the legendary jazz saxophonist as the world knows him, but John Coltrane as an incarnated divine musician creating sacred and transformative soundscapes. Through the efforts of some devotees, he was later proclaimed a "saint," and was embedded in a dedicated Christian church affiliation that centers on him and his music. This fusion of jazz and religion is not as strange as it may seem at first. A long-standing connection exists between music—jazz in particular—and religion, and their commonalities and mutual links have been noted by various scholars (Leonard 1987; Sylvan 2002; Bivins 2015). Some writers even claim that religion is the primary source for jazz music (see Cox 1993; Floyd 1996; Stowe 2010). In the case of the Saint John Coltrane Church in San Francisco, the connection seems an apparent Christian one, although the history of the church and the diversity of its practices and membership complicate such a simple equation. Our research focuses on the extent to which Coltrane's music is to be regarded as a self-standing spiritualizing force or movement, an implicit form of religiosity, and how the movement has been influenced by the urban context in which it is created and performed.

Jazz, religion, and the urban

Before going into detail about the religious dimensions of Coltrane and his music, the reputed link between jazz and religion requires some explanation, as it is usually not self-evident. To understand this relationship, we briefly examine the origins of jazz during the times of the struggle for emancipation by African Americans until its incorporation into subjectified present-day religious or spiritual cultures.

Jazz initially became a vehicle for solace and hope in the midst of racism and oppression. As jazz theorist John Gennari notes:

> Jazz was forged in the cauldron of Jim Crow segregation by the descendants of slaves, who transformed antebellum spirituals, work songs, hollers, and ring shouts into the witness-bearing, intensely expressive truthfulness of the blues, as well as the effervescent spirit of ragtime. (Gennari 2005: 1167; cf. Peretti 1992: 14–21)

These conditions are usually taken as the biotope for the connection of jazz and religion, and scholars such as Harvey Cox, David Stowe and Jason Bivins have pointed out that jazz always has been entangled with religion and spiritual eclecticism, reflecting the influences of gospel, folk spirituals, the "black church" and other sacred music traditions (Cox 1995: 139–57; Stowe 2010; Bivins 2015).[2] This relationship is particularly evident in Pentecostalism that, as a form of revivalism with an important emphasis on ecstatic expression, became the primary religious practice for many African Americans, its musical traditions and expressive styles intersecting with the development of jazz. Both Pentecostalism and jazz, as eclectic vernacular movements, were born in heterogeneous urban environments at the beginning of the twentieth century (Pentecostalism in Los Angeles and jazz in New Orleans), and were similarly condemned and ridiculed at the time. As Cox asserts, both jazz and Pentecostalism "sprang from the same womb," emerging from the African American experience of resisting oppression through enthusiastic worship and performative expression, and there are parallels between possession and trance states, speaking in tongues, and forms of jazz singing and performance (Cox 1995: 145–6; Cox 1993). Furthermore, both jazz and Pentecostalism are characterized by a synthesis of expressive styles, and neither would have developed if not for the open, free-spirited atmosphere and the cultural and racial intermingling of the urban milieu in which they were practiced. Jazz began as a uniquely urban phenomenon and was able to expand and flourish because it developed in these urban locales and spread throughout the American cityscape (New Orleans, Kansas City, Chicago, Memphis, St. Louis, New York,

Detroit, Los Angeles, etc.). The growth of jazz also coincided with the migration of millions of African Americans who moved from rural areas and sought employment in the cities at a time when there was increased demand for industrial labor. It was in the cities where black musicians found each other and where an urban culture existed that allowed for the exposure to and experimentation with varied musical styles, resulting in the emergence of new musical communities, as well as affording the venues and the necessary audiences for the "counter" communitarianism that became jazz.[3]

Like rock'n'roll and other forms of popular music, jazz was from its beginnings often depicted as the antithesis of religion, expressing and evoking the hedonistic pleasures of Saturday night in contrast to the holiness of Sunday morning worship services (Murray 1989; Stowe 2010: 312–13), a notion that reflects an overly simplistic binary concept, analogous with the Bakhtinian idea of inversion and the functionality of a carnivalesque Shrove Tuesday celebration prior to the austere religious cycle of Lent. In reality, the religious and secular expressions of jazz were and are intertwined, influencing and riffing on each other, as musicians and audiences often embraced both contexts equally. As Zora Neale Hurston observed in her WPA folklore documentation in the 1930s, the spirited jazz played at parties and at sanctified church services, sometimes by the same performers, often were musically indistinguishable (Hurston cited in Levine 1977: 180). The emergence of jazz, regarded as a quintessentially American and archetypically modern art form, is also celebrated as the first major black American art form, infused with the religious traditions and rural vernacular culture of African Americans, creatively transformed into a music of the city, an expression of urbanization and social change, resistance and freedom, the power of improvisation and pluralistic sensibilities, the vitality of urban life, and the desires of modernity.[4]

Research on the relationship between jazz and religion generally focuses on the following four variations of this connection: (1) religion as an inspirational source for jazz musicians; (2) jazz music composed and written for religious goals or having a (historical) religious context (for example, gospel music); (3) jazz as musical expression functioning or used within formal religion or religious contexts (performing jazz musicians), and (4) jazz music and/or its performer as having a sacred stature as a religious phenomenon in and of itself (cf. Stowe 2010: 313–14). In the academic literature on jazz, such categorizations are sometimes obfuscated by the omnipresent insertion of religious metaphors: "jazz as a religion" or jazz music or jazz musicians depicted as a sect or cult, with the musicians referred to as "high priests" or "prophets," professing "myths," singing "sacred texts" and practicing "rituals," whether directly in the presence of their devotees or not.[5] In this chapter, we would like to move beyond such metaphorical descriptions, as well as the notion of jazz as a surrogate religion, and explore by means of ethnographic data the religious meanings of the music performed and experienced in the Saint John Coltrane Church. This is a performance of jazz that appears unique in its expression and that needs to be understood and researched emically and in its own way, apart from the previously mentioned categorizations of jazz and religion. Although the Saint John Coltrane Church is one of the few affiliated churches in the world that includes jazz as a central part of its formal services, we explore the ways that the church also may be a framework to allow Coltrane's music to independently perform its salvational and transformative meaning. As a hybrid and alternative religious phenomenon, how is the church's musical expression particularly viable in modern urbanity, and what is its enduring appeal and meaning for congregants and visiting jazz enthusiasts? Does the "urban magic," as Bivins (2015: 145–6) puts it (a specificity of the urban musical and social communitarianism that enabled jazz), work in this context for creating new modes of sociality and religious experience? The broader issue of jazz as an interface between religion and the urban is hence explored in our essay: in particular, to what degree has the forced relocation of the Coltrane Church and the urban renewal and gentrification of the San Francisco Western Addition district where the church was originally situated affected this religious movement and ongoing formation, and how has this specific expression of jazz been shaped, influenced, or stimulated by the urban habitat in which the Coltrane movement arose and evolved?

Urban renewal and the Coltrane movement

The cityscape of San Francisco has been rapidly changing during the last two decades and this transformation has impacted the Coltrane Church. The economic effect of the high-tech businesses of "Silicon Valley" has accelerated urban renewal and a massive construction boom has transformed the city and the entire Bay Area. Researcher Rebecca Solnit documents the impact of this urban makeover on San Francisco's various districts in her book *Hollow City: The Siege of San Francisco and the Crisis of American Urbanism* (2000). From her at times nostalgic and moralistic perspective, she describes the city's urban outward change. She points out how old neighborhoods and communities that were once "authentic" and unique to the city have become "hollow" and soullessly "siliconial" through gentrification and urbanization. Whether or not one agrees with Solnit's judgment, this transformation has strongly affected the Western Addition and Mission districts of San Francisco. The result of urban renewal and the applied architecture that one now finds there is often of a rather prosaic kind, with relatively low buildings of an unassuming style in clean and neatly renovated blocks, reflecting new forms of habitation and social structures. There is little left here that refers to the historic blocks where a vibrant African American community known as the "Harlem of the West" once thrived as a center of black culture—its barely visible remnants now marked by plaques scattered on the sidewalk. The bland redevelopment of the area is exemplified by the modern low-rise building near the corner of Fillmore and Eddy Streets, not the setting one would connect to an area previously famous for its jazz scene and nightlife. It is here that Saint John Coltrane Church was situated until 2016, in a drab annex to the West Bay Community and Conference Center.[6]

Figure 10.1 The modern John Coltrane storefront church at Fillmore. A poster with crossed saxophones is displayed to attract the attention of passers-by, August 2011. Photo by Peter Jan Margry.

The church was not always located here. Formalized in 1969 by hairdresser Franzo Wayne King[7] two years after Coltrane's death, the movement began in King's apartment in a Potrero Hill housing project in 1967 and then relocated to an after-hours jazz club in the Bayview-Hunters Point district, where part of an apartment was converted into a chapel in which people could gather at midnight once a week and devote the next 24 hours to prayer, fasting and sound immersion by listening to Coltrane's album *A Love Supreme* (Boulware 2000). Realizing he needed more room for those devoted to Coltrane's music, in 1971, King and his wife Marina rented space in a commercial building in the Fillmore neighborhood and transformed it into a storefront "church," a phenomenon common in poor African American urban communities. The location was only a few blocks away from Haight-Ashbury, in a historically black district developed by African Americans who emigrated there in the 1940s to escape racial segregation laws and to take shipbuilding jobs during World War II. During the 1950s and 1960s, the Western Addition was the epicenter for a brutal urban redevelopment plan that demolished much of area's historic architecture, displaced thousands of residents, and decimated the community and local black culture.

For nearly three decades, the Coltrane Church was located in a surviving historic Victorian-style structure on Divisadero Street (no. 351), a popular and well-known venue, due in part to its social programming and activist work, with space for community participation, free hot meals and food programs (some in collaboration with the Black Panther Party in the 1970s), shelter and clothing for the homeless as well as counseling, music lessons and other types of community outreach. Over the years, the church developed into a community institution, providing an impetus to the neighborhood and offering opportunities to the poor and needy, as one of the last remaining grass-roots African American jazz venues in San Francisco and as a community-based institution committed to issues of social justice and local outreach. As King explains it:

> We have been a vibrant part of this community since 1971 in large part due to our social activism; speaking out against violence, racism, police brutality, and other important issues of equality. We are here in this community as a spokesperson that speaks truth to power. (MacDonald 2015)

On many Sundays, the storefront space was packed, the enthusiastic congregation of musical devotees sometimes overflowing onto the sidewalk. However, in the year 2000, rent increases pushed the church out of the neighborhood.[8] This expulsion was a consequence of the social and cultural changes in the Western Addition district, the result of a second wave of urban renewal and gentrification that began in the 1990s (an ongoing process of gentrification and "de-race-ification" referred to as the "negro removal" by some local residents; see Solnit 2000: 43; Klein 2008: 6). Capitalizing on the strong economy and the dot-com and commercial real estate boom, high-tech firms, social network companies and other businesses persistently invaded the area, purchasing property, raising rents, evicting tenants and making housing too expensive for many of its predominantly African American and Latino residents, wiping out what remained of the former "character" and ethnic cultures of the old district. The overall gentrification of the area and the eviction of the Coltrane Church not only upset community members and cultural critics such as Solnit, but even the city's mayor Willie Brown expressed his regrets about the relocation of the church, stating, "Do I care? Yes, I do. I wish I could have it like it was, many years ago" (Brown cited in Boulware 2000). The changing demographics of San Francisco are indicated by US Census Bureau figures that reveal among other things the dramatic decline of the city's African American population, showing a steady decrease in the community from 13.4 percent in 1970 to 10.9 percent in 1990, and then shrinking further to 6 percent by 2013, a reduction of nearly 60 percent since the 1970s.[9]

These changes in demographics and the ongoing gentrification of the neighborhood did not leave the Coltrane movement unaffected, and after the church moved to an affordable but more distant site, local attendance declined further. By the time the church relocated in 2007 to a storefront space in the heart of the Fillmore district, even more long-term local residents had been forced to move away because of rent increases and evictions.[10] In 2015, Reverend King expressed his dismay about the exodus of the African American community and the loss of local culture: "What's happening is just short of genocide. It's a crime, premeditated with malice and forethought, to drive the African-American community out of San Francisco. We've tried to stand up against this fact with very little success" (MacDonald 2015). Even today's visitors to the Coltrane Church have commented on the effects of these demographic changes. For instance, Hanna Morjan, from Scotland, recalled her experience at a Sunday service in April 2015:

> I have very mixed feelings, the music was great and being invited to join and given a tambourine to play helped me to feel part of it, to feel the music in my body. But I was disappointed by the lack of people, and hardly any African Americans. . . .[11]

Ironically, the Fillmore and Western Addition areas are now promoted and advertised as "the chic and trendy Fillmore Jazz Preservation District of San Francisco's classic Western Addition/Alamo Square area."[12] This recent sales pitch and branding of the neighborhood further reflects how this once distinctively African American district is being marketed and reenacted in a safe and heritagized way, with an increasingly gentrified, tourist and cool yuppie vibe. These changes have affected attendance at the church, as those in audience shifts from regular members who once were from the local community to growing numbers of visitors and tourists. In this way, urban gentrification has disrupted and decreased the local community involvement, and as more outsiders and jazz tourists visit, the audience and context changes from a "Church" experience for regular weekly devotees into a more implicit expression of religiosity for one-time visitors and jazz aficionados for whom the performance of Coltrane's music alone provides the transformational experience. Or, as Coltrane expressed it himself: the experience of the divine *realized* through sound.

Sanctifying John Coltrane

Time and again, the relationship between jazz and religion is exemplified by the musical and spiritual legacy of John William Coltrane (1926–67). Considered one of the single most influential jazz musicians of the twentieth century, Coltrane is revered by many as part of a "holy trinity" of jazz performers—a religious metaphor anew—alongside Louis Armstrong and Charlie Parker, Duke Ellington or Miles Davis, depending on one's favored consecrated trio. While Ellington and Mary Lou Williams are acknowledged for creating new sacred music, and Sun Ra's cosmic Afrofuturistic mysticism inspired a cult following, it is Coltrane and his musical legacy that over time have been most frequently and reverentially situated in a wider spiritual domain (cf. Brown 2010; Nisenson 1995; Saul 2003: 254–60; Whyton 2013: 60–68; Woideck 1998: 47–8, 66–7). Reflecting on the religious representations and enduring adoration of Coltrane, writer Francis Davis proclaims "More than any other performer of his time or ours, he is a god we create, if not in our own image, then according to our desires and beliefs"

(Davis 2001: 33). Coltrane's own religious awakening occurred in 1957, at a time when he was struggling with alcohol and heroin addiction, and the experience transformed his life and musical aesthetic. As he later wrote in the liner notes for *A Love Supreme*:

> I experienced, by the grace of God, a spiritual awakening, which was to lead me to a richer, fuller, more productive life. At that time, in gratitude, I humbly asked to be given the means and privilege to make others happy through music.[13]

While Coltrane was raised in the African Methodist Episcopal Zion tradition, he later adopted a broad, non-sectarian eclectic view of the divine that was accepting of all faiths and he viewed his music as an expression of a universalist spirituality: "I'd like to point out to people the divine in a musical language that transcends words. I want to speak to their souls" (Coltrane cited in Porter 1998: 232). As a result of the enormous success of his album *A Love Supreme* (recorded in December 1964 and released in 1965), his concept of multicultural musical transcendence became extremely popular in the late 1960s. His early adoption of Eastern spiritualities is evident in recordings such as *Om* (1965) and *Meditations* (1966), while his *Ascension* album (1966) shows his move towards the incantatory and the shamanistic, bringing in the "magical powers of repetition" (Santoro 1992: 499). His second wife Alice played an active role in the ways Coltrane started to explore non-Western spiritualities and to look for his "authentic self," as he was influenced by the Bhagavad Gita, Theosophical texts, the *Tibetan Book of the Dead*, Krishnamurti, Yogananda, the Kabbalah, yoga, astrology and Sufi mysticism; he was also meditating daily and began experimenting with LSD in 1965 (Berkman 2007: 44–5, 55; Nisenson 1995: 166–7). The spiritual universality he wished to realize through music was an intentional path to religious pluralism and away from the institutionalized religious traditions, an attempt to use jazz music as a universal vehicle for modern spiritual self-realization (Berkman 2007: 56). The album *Universal Consciousness* released in 1971 by his widow Alice Coltrane after his death in 1967 is a later example of that spiritual quest.

As a result of the emotive power and spiritual status of his music, many Coltrane enthusiasts consistently have transposed him from the secular to a sacral realm. Why and how did this happen? Was it due primarily to his own deep spiritual commitment and pursuit of God, or the legendary accounts of him having a charismatic and "holy" presence when he performed? Or is this sacralization due to his pluralistic and universalist approach to religiosity or to an "independent," transformative agency of his music in particular that evokes transcendent and sublime experiences of the divine?

In the case of the Coltrane Church, the movement was inspired and influenced by all of the above. When the church initially developed in the late 1960s and early 1970s, the small congregation adopted various names, including the "One Mind Temple Evolutionary Transitional Body of Christ," reflecting the belief that Coltrane was an evolved manifestation of Christ and that all of humanity is of one mind, one connected consciousness, with the potential to evolve into Christ-like beings as well, guided by Coltrane's music and wisdom.[14] At that time, King (who for a while called himself Ramakrishna Haqq) and his congregation worshiped Coltrane as "an earthly incarnation of God" and a second coming of Christ, but also considered him a manifestation of Hindu Lord Krishna ("the enchanted player of the flute"), as they studied Vedic scriptures and sacred texts from various religious traditions, and were influenced by black liberation theology and their interactions with Alice Coltrane and her devotion to the popular Indian guru Sathya Sai Baba. Vestiges of these influences are still evident on the church's website, which refers to the "mighty mystic" Coltrane as Sri Rama Ohnedaruth, the Hindu spiritual name given to him by Alice Coltrane after his death.[15] During this time of spiritual exploration in the 1970s, King's One Mind Temple was located a few blocks away from another temple, the Jim Jones Peoples Temple. After the Peoples Temple relocated to Jonestown (Guyana) and ended there in 1978 in an apocalyptic mass suicide, various alternative religious movements flourishing in San Francisco were scrutinized and condemned. The Coltrane temple also was seen as "alternative" and criticized by some for its "cult" worship of Coltrane (Boulware 2000). Indeed, scholars have argued that since the 1950s, San Francisco became known for the "weakness" of its traditional churches, and that became in some ways a social laboratory, establishing an "authentic" open urban platform for the development of new religious movements and "cults" (see Stark and Bainbridge 1981).

Around the same time of the trauma of Jonestown, Alice Coltrane sued the church for $7.5 million, accusing it of exploiting her husband's name and violating copyright laws; the case received national attention after the *San Francisco Chronicle* covered the dispute with the headline "Widow of 'God' Sues Church" (Boulware 2000). Amid this controversy and under increased scrutiny, the Coltrane Church was approached by members of the African Orthodox Church, a small denomination searching for new membership and the expansion of their fledgling organization. In response to these overtures and offers of support, King pursued a somewhat more legitimate status in religious territory and in 1982, the previously informal or "lay" Coltrane movement was incorporated into the African Orthodox Church. Coltrane was subsequently canonized on September 19, 1982 by Archbishop George Duncan Hinkson and called Saint John.[16] Coltrane was not the first saint in the AOC, as he joined other important African American political leaders and civil rights activists such as Marcus Garvey and Martin Luther King, Jr. who were canonized before him. It is impossible to know how the humble Coltrane would have felt about this sanctification, but in his quest to live a truly religious life, sainthood was a status he alluded to in 1966: when asked by an interviewer what he would like to be in the future, he replied, "I would like to be a saint" (Porter 1998: 260).

Within the formal exegesis of the Saint John Coltrane Church, the consecrated Coltrane is often addressed with his two first names as "Saint John Will-I-Am," which is a reference to Exodus 3:14 in which God says to Moses from the burning bush: "I AM THAT I AM: and he said, Thus shalt thou say unto the children of Israel, I AM hath sent me unto you." The nicknaming of Coltrane as the "Risen Trane," is a reference to the transformed John Coltrane "post-1957," after he emerged from drug and alcohol addiction, and points equally to a specific Christian context—on the one hand to Coltrane's resurrectional success as a person overcoming human vices, while at the same time suggesting an analogy with the resurrected Christ. However, in line with the doctrines of African Orthodox Church, as Archbishop King explains, Coltrane has been accommodated to the formal Christian teachings and downgraded to a saintly figure, not a prime godly figure: "We demoted Coltrane from being God. But the agreement was that he could come into sainthood and be the patron of our church" (King quoted in Freedman 2007). Yet, the sacred and godly status of Coltrane still seems prominent, a synthesis of beliefs, as explained on the Church's Facebook webpage in a somewhat hybridic manner:

> Our primary mission . . . is to bring souls to Christ; to know sound as the preexisting wisdom of God, and to understand the divine nature of our patron saint in terms of his ascension as a high soul into one-ness with God through sound.[17]

In this way Coltrane is referred to both as a divine and ascended godly person and as a saintly mediator to bring people to Christ, with Coltrane's musical sound described as a direct expression of God, even as the "pre-existing wisdom" of God.

Throughout its 45-year history, the Coltrane movement has been led by Franzo King and his wife Marina King as pastors and church founders, who later were appointed Archbishop and Reverend Mother within the context of the African Orthodox Church. Numerous other individuals have been appointed reverends, deacons and sub-deacons, including the King family members Wanika King-Stephens, Makeda King Nueckel, Franzo Wayne King, Jr., and Marlee-I Mystic, among others. Musicians in the church ensemble Ohnedaruth (Sanskrit for "compassion" and as mentioned, one of Coltrane's spiritual names) have been appointed reverends and clergy as well. There are various other members of the ministry who assist in the services and participate in the jam sessions, including two ministers of tap dance. According to King, this church "is born out of music, a gift of God," working to "strip down the dogma" and "bring the people in an enlightened state, to a love supreme."[18] As the group is now established formally as the Saint John Will-I-Am Coltrane African Orthodox Church, God is worshipped through sound and saintly Coltrane is honored by them as an enlightened being who attained mystical union with God and conveyed this through his music. In this context, improvised jazz itself becomes the vehicle for transformative experiences of the sacred. Coltrane's compositions, venerated as holy and divinely inspired, have been elevated to a level of sacred song conventionally reserved for more traditional forms of devotional music. Officially, the primary principle of the church is now defined and limited to the worshipping of the Christian God, with a mission to "bring souls to Christ."[19] Or as King once phrased it, "when you listen to John Coltrane you become a disciple of the anointed God" (Freedman 2007).

But this assertion begs the question: who now attends the services and how connected is the audience to the official ("Christian") rhetoric, and what do people profess to experience there? Do the musical experiences of the increasing numbers of new visitors and jazz enthusiasts supersede the formal liturgy and instead connect to the perceived and experiential sacred power of Coltrane's music as expressed earlier by the Coltrane movement prior to its formalization as part of the AOC in 1982? As a case in point, the description provided by Oriah Vaughn, quoted at the opening of this article, shows an eclectic religious-spiritual stance that is familiar for a person of her background in the present day, and one that seems to connect clearly to the universalist Coltrane spiritual setting as it originally emerged, before its AOC affiliation. Characterizing her approach to non-institutional spirituality and her inclination toward Coltrane's jazz incantations, Oriah further explains:

> I pick and choose my personal beliefs from a variety of religions. It's more of a synthesis of "truths" drawn from a broad base. I believe every religion has something to teach us. Religions are meant to be dynamic and incorporate new knowledge; unfortunately, they don't get updated too often. I think a lot of times people say they are "spiritual, but not religious" to cover the fact they don't have a strong or well developed system of beliefs that ground their spiritual practice.[20]

Oriah's views clearly express an approach to religion that is widespread and meaningful for many individuals outside an institutional religious framing, who seek and adopt other forms of spirituality.

For those who visit the church and are immersed in Coltrane's sublime soundscape, what are the possible religious contexts and ways to consider the personal spiritual encounters and experiences that occur? More broadly, how does religion manifest itself in other forms and expressions in addition to the obvious churched forms? In pursuit of such questions, we take the approach that religious and spiritual experiences need to be understood synonymously as common and culturally created expressions of religiosity. In this regard, the music of John Coltrane itself also may be considered a religious phenomenon and be regarded as an expression of "implicit religion."[21]

Jamming jazz as sacred soundscape

From the street, St. John's is easy to miss, situated in a plain office building, its nondescript glass front door looking like an entrance to a business. The church is identifiable only by a window poster of a cross formed from two tenor saxophones and a small sign that says "Coltrane Lives," although on the days of worship services, a larger sidewalk sign is put out on the street. The church is housed in a simple and rather small space that may contain up to fifty people, with rows of blue banquet chairs facing the combined stage/altar that is cluttered with a full drum set, keyboards, a standup bass, saxophones, amplifiers, microphone stands and other instruments. The walls are decorated with large colorful Eastern Orthodox-style icons created by church Deacon Mark Dukes, with images of the tree of life, fiery-winged red angelic beings, the Blessed Virgin Mary and child, and a dreadlocked Jesus sitting on a throne, all depicted as dark-skinned in the aesthetic tradition of the African Orthodox Church.[22]

To the left of the altar is an eight-foot tall image of patron saint Coltrane sitting on an African throne, wearing white religious vestments, framed with a golden halo, holding a saxophone blowing holy fire and a scroll with words from the liner notes of *A Love Supreme*. There are batik depictions of Coltrane and assorted African motifs hanging from banners along the ceiling and against the walls, and an image of Che Guevara prominently displayed on a conga drum. A small table in the back of the church contains a guestbook, a pamphlet with the title "Are You an Addict?" and a few items for sale, such as Coltrane Church T-shirts, incense,

Figure 10.2 Icon painting of John Coltrane entitled "Saint John the Divine Sound Baptist" by Deacon Mark C. E. Dukes, ca. 1989. This is a frequently reproduced icon of Coltrane, with his tenor saxophone ablaze with the flames of the Holy Ghost, while the scroll in his right hand quotes the liner notes from *A Love Supreme* and the lyrics to the fourth movement. Photo by Dan Wojcik.

prayer cloths and icon postcards. Hanging above the table is a frequently reproduced icon of Coltrane—holy, intense and otherworldly in expression—in a green velvet jacket, gripping his flaming saxophone.

The brightly lit converted office space has no resemblance to a conventional jazz venue as it might be experienced now or as imagined in the jazz clubs of yore—this is no dimly-lit, smoke-filled, lush life refuge "where one relaxes on the axis of the wheel of life, to get the feel of life, from jazz and cocktails..[23] Within the newly gentrified, renovated district, this open-door sanctuary frames jazz within the modern Coltrane paradigm of a global, more gentrified audience, a sacred space that is presented in an accessible way to be easily experienced by all those longing to be taken away on the musical waves of Coltrane's music.

The worship services at St. John's are scheduled to begin at noon (to accommodate the late-night schedules of musicians), but rarely start on time. Prior to the service, the atmosphere inside the church is relaxed, as congregants socialize, the musicians set up and then warm up, and the international visitors and tourists trickle in, usually sitting towards the back of the room and snapping photos until a church member walks over and politely tells them that photography is not allowed. Upon entering the storefront space, the visitors are often greeted at the door by one of the parishioners, or sometimes welcomed by Archbishop King himself, his gold tenor sax and the gold cross hanging from his neck competing in religious symbology. The core congregation is made up of a mix of young and elderly African Americans and a few white musicians and multi-ethnic parishioners, many casually dressed, a few in suits or African attire, some of whom bring their entire family to the service, and children freely wander about the room and participate in the service. These regular members are joined each Sunday by a potpourri of jazz enthusiasts, hipsters, spiritual pilgrims, curious locals and travelers from all over the world who have heard of the worship services, known and promoted by enthusiasts globally for their lively style and enthusiastic performances of Coltrane's music.[24] Marlee-I Mystic, a deacon in the church, informed us that weekly attendance is typically somewhere between ten and twenty people, including local and international visitors. During our fieldwork at the church,[25] we observed a similar number of those present, with the church's musicians, choir and regular members sometimes outnumbering others in attendance, many of whom were travelers and Coltrane devotees from afar.[26]

Whatever their religious intention, regular church member or not, those participating at St. John's are encouraged to bring their own instruments to contribute to the service, and members of the audience are handed tambourines and asked to participate as the spirit moves them, with dancing and personal "witnessing" equally encouraged as well. The church's ensemble, Ohnedaruth (also referred to as the "Ministers of Sound") and a small choir called the "Voices of Compassion" (formerly the "Sisters of Compassion") lead the congregants in what they call the "Coltrane Liturgy," which begins with a nearly two-hour jam session that combines the liturgy of the African Orthodox Church with the harmonies, melodies and rhythms of Coltrane's musical sermon or "prayer," *A Love Supreme*, as well as other Coltrane works.[27] On the occasions that we attended the services, Coltrane's "Africa" was played during the processional and the soulful ballad "Lonnie's Lament" performed during the Introit. When the ensemble played "Acknowledgment" from *A Love Supreme*, the choir sang the words to Psalm 23 ("The Lord is my shepherd . . . "), and those present were encouraged to say (or pray) the words "A Love Supreme" at the appropriate moment in the performance, with the core congregation and visitors alike chanting in unison. In a similar synchronization of formal liturgy with Coltrane's music, the choir sang The Lord's Prayer when the composition "Spiritual" was played. After the musical performance, additional traditional Christian liturgical elements are introduced, such as readings from the Epistles and the Gospels, the Apostles' Creed, the offering, and then the sermon. While the church follows in a formal way the creeds of the African Orthodox Church (a blend of Eastern and Western liturgies and traditional Catholic doctrine),[28] the services also are strongly influenced by Pentecostalism, with an emphasis on the presence of the Holy Spirit, spontaneous shouting-out, clapping, the exorcizing of demons (through music), and Archbishop King's own fiery preaching. After the service, as the musicians pack up their instruments, Archbishop King and other church members mingle, shaking hands and chatting with those still in the audience, the friendly and informal mood similar to having just participated in a jam session in someone's living room. On one occasion, homemade pumpkin pie was shared with everyone, and another time, Archbishop King, who observed that we had stayed for the entire three-hour jazz mass, walked over, greeted us, and in his familiar down-to-earth manner, laughingly said, "You cats ate the whole thing!!"

In 2015, at 70 years of age, the Archbishop King continues to guide the Coltrane worship as he has for more than 45 years, with a welcoming style and openness that accepts all who attend the services. As an accomplished saxophone player, each Sunday he energetically performs his religious calling, playing his horn much of the time but also preaching, sometimes drenched in sweat after an hour or two of passionate saxophone praise for St. Coltrane. The Pentecostal roots that he embraces infuse the performance, and each Sunday event is improvisational in its orchestration, flowing in the way that the spirit moves him and other members, and infused with his jazz expertise and aesthetic: he dances slowly with his saxophone or with Mother Marina and other members of the swaying, full-voiced choir; he blasts his horn hard with ecstatic abandon to create elevated squalls of sound, and then slides smoothly into soft solos of soulful worship. Throughout the ceremony, he moves about the room in tune with other members of the congregation and in sync with the holy noise, shouting out in tribute or holding his hand over the head of another enraptured soloist as if to feel the ascension of the sacred harmonies, raising his arm high to the sky and pointing to the heavens in rhythm with the hallowed sounds. King is accompanied by equally enthusiastic performers, who sacrifice themselves on the altar of sound praise to produce illuminated layers of inspired music each Sunday, with some musicians dropping in for a portion of the service, while others are present for the duration, with regulars Reverend Sister Wanika ever elegant on the upright and electric bass and Reverend Max Ha'qq on alto sax, blowing with such furious intensity that he looks as though he might explode. As the musicians and choir pour their souls into the creation of the exalted tones, jamming in lengthy and unrestrained solos, parishioners sway with the music, some dancing in the aisles, clapping or clasping hands together in prayer, and shouting "Amen!," "That's right!" and "Hallelujah!." Tambourines in hand, throughout the service many of those present participated in this spiritual Coltrane jazz fest, as the friendly, communal vibe welcomes and includes everyone, curious visitors and tourists, ethnographic observers, jazz pilgrims and longstanding devotees alike, all grooving in tune to the sacred sounds.

While the later portion of the service is highlighted by the sermons of Archbishop King or the Reverend King-Stephens, the Coltrane Church experience is dominated by the music and its claim to redemptive power, emphasized and performed almost non-stop for nearly two hours. Describing the worship services, participant and local resident Kevin K. stated that "The first two hours or so are taken up with what the pastor calls 'exorcism of the demons,' but to me are some of the most intimate and uninhibited jazz jams I've ever encountered" (2006).[29] This observation, supported by our fieldwork and the comments of other visitors, touches directly upon the question we pose here. What is the "church" in its institutionalized forms and what is the Coltrane musical spirituality in its own "implicit" expression, as an independent and non-churched experience? In other words: is this religious expression to be seen as an idiosyncratic but formal Christian parish, or is it better understood as a religious movement based on Coltrane's musical aesthetics and sacral sound, and only facilitated within the formal framework of a "sectarian" branch of the small African Orthodox Church? As explained above, the eclectic and grass-roots St. John Coltrane movement started outside of any formal Christian institution and it was only in 1982 that external pressures compelled King to join the African Orthodox Church and pursue training in order to be ordained a bishop by the AOC.[30] As a result of this formal Christian incorporation, the idea that Coltrane himself was God had to be seen a fallacy and King reluctantly accepted a demotion of Coltrane into "just" sainthood in order for him to continue to be patron of the church (Washington 2001: 394; Freedman 2007).

In a related manner, this particular St. Coltrane branch of the AOC is strongly independent and conveys a double message: the Christian and Coltranist. And the mission of the Coltrane Church remains international in scope: "to paint the globe with the message of *A Love Supreme*, and in doing so promote global unity, peace on earth, and knowledge of the one true living God."[31] While this mission is presented within its orthodox Christian dogma of monotheism, it also promotes a pluralistic and holistic spiritual dimension that also integrates curious tourists, jazz devotees and individuals from other religious traditions into its broader sense of purpose. As Archbishop King proclaimed recently in one sermon we attended, "We are part of the African Orthodox Church, but we [Coltranists] are a universal church, a revolutionary church," and the sermons by King and Reverend King-Stephens repeatedly emphasized the church's activist and inclusive nature, with references to Buddhism, Hinduism, Bob Marley, the Dalai Lama, Plato, Martin Luther King, Jr. and Coltrane's own personal search for "religious truth" beyond one particular spiritual tradition.

In addition to the inclusive and interfaith subject matter of the sermons, the weekly services at the Fillmore parish are distinctly different from that of a "regular" or mainstream Episcopalian or Catholic Orthodox service, a sharp contrast indeed as the Coltrane Church's informal and improvisational style is laced with Pentecostal gusto and spontaneity, as congregants and visiting tourists dance, shout out, play instruments and interact with the ministry in a jazz-infused and close-knit context. In June 2012, Kyle M. from San Francisco gave a more or less typical depiction of how spiritual and "unchurched" visitors perceive the service and participants present:

> The Church of St. Coltrane, though, is all about acceptance. And when I say acceptance, I mean EVERYONE is welcome . . . Basically, the place was lightly organized chaos. Perhaps that's why I didn't meet any regulars. Everyone who came to church was coming for the first time. Most were foreign. . . .[32]

A similar ratio of those in attendance also came to the fore when on a Sunday in January 2000, one of the church leaders asked the audience how many of them were local residents. Of the approximately sixty people present, just three raised their hands, with more than 90 percent of the congregation from somewhere else, with visitors from Texas, Arizona, France, Spain, New Zealand, Denmark, Sweden and Ireland (Boulware 2006). Similarly, every time we attended services, Archbishop King made a point during the sermon to ask people where they were from and it was clear that there were always more travelers than regular congregants. The number of formal members of the Coltrane Church is in fact relatively few, with the core congregation varying from 15 to 25 parishioners over the years.

Still, the weekly audiences not only show a global geographical representation of jazz lovers and votaries, but a wide range of believers from assorted religious and spiritual currents and agnostics as well. Although now formatted in the frame of the AOC, the Archbishop King and other members of the church simultaneously put forward an open spiritual paradigm, in a way that almost any religiously inspired person or lover of jazz may be accommodated. In an interview in 1999, King reflected on the expansive spiritual dimension of Coltrane and his meanings: "I realized that the music of John Coltrane was representative beyond culture . . . And it wasn't just a cultural or ethnic thing. It was something that was higher." King views the church and particularly the soundscape of St. John's as the "genesis" of an autonomous Coltrane belief system. As he states it, at some moment

> . . . you begin to see God in the sound. It's a point of revelation, it's not something that happens with absolute clarity, but it begins an evolution, or

> a transition, or process. The consciousness level, that opening, is evolving. Baptism is what it is.[33]

This "baptism in sound" clearly plays with the other baptizing John in Christianity, but refers here to the possibility of Coltrane's music to touch and capture the hearts and minds of the listeners and to realize transformation.[34] And so, at St. Coltrane's storefront church two realities of religious experience can be perceived at the same location, representing actually a *simultanaeum* of the formal services of the African Orthodox Church and at the same time the open spiritual Coltranist domain that provides the experience of the musical sublime, creating spiritualizing effects among the majority of visitors who know nothing about the doctrines of the AOC.

Franzo and Marina King's own personal "sound baptism" and spiritual transformation occurred in 1965 during a live performance by Coltrane at a popular club in San Francisco that changed their lives forever, an experience that they equated with the presence of the Holy Ghost, filling their hearts with the love of God. Other individuals have described comparable transformative experiences occurring because of Coltrane's music. For instance, saxophonist Robert Haven (aka Roberto DeHaven), who became minister in the church, states that

> For me, Coltrane had this very powerful influence in that he was like me, he was using heroin and drinking, but then he quit. Then he went on to devote his music to God. I would sit in my room and cry listening to Coltrane solos . . . I was completely under Coltrane's spell.

For congregant Jon Ingle, Coltrane's music restored his religious belief:

> I grew up in Texas, and for a long time I had this little war going on with God . . . I turned away from myself and my spirit. John Coltrane has led me back. So I feel like the spirit of John Coltrane has led me to being more fulfilled in my life than I ever could have imagined.[35]

Franzo King is clear about Coltrane's non-denominational and open sacral and spiritual qualities:

> We don't hold a monopoly on John Coltrane. John is a saint among Buddhists; he is a saint among Moslems. He is a saint among Jews. And I think there are even a few atheists who are leaning on that anointed sound.[36]

And so the church is a shared locus for both views, as King himself recognizes that there are indeed different religious expressions present: the non-mainstream but more or less formalized African Orthodox Church, and the open and more "implicit" Coltranist musical expressions perceived by non-AOC-churched Coltrane devotees from all over the world who are able to have transformative experiences brought up solely by the performance of Coltrane's music.

That open spiritual domain connects well to what Coltrane himself professed about religiosity in general: "I believe in all religions."[37] As a spiritual seeker, his religious eclecticism was related to his conviction that music had the power to expand consciousness in positive ways and transform people as a force for good, and that the divine could be more strongly expressed through music than through words. In this spirit, the Coltrane movement embraces a tolerant and respectful pluralism, based in part on Coltrane's own religious awakening and expansive approach to spirituality and also due to the religious diversity of its congregation and that of San Francisco's social variety and the visitors attracted to the city year-round. As Deacon Marlee-I Mystic characterizes it:

> The Coltrane Church's approach to spiritual expression allows for a New Age-style of fellowship that is democratic and non-proselytizing, using the universal language of music as a platform to praise, meditate and fellowship with a galaxy of kindred souls from all walks of life . . . and fashions a home for Bay area residents and the world community of travelers seeking a cosmic religious experience.[38]

Do the beliefs and spiritualities associated with John Coltrane represent then one sort of future for religion in the urban sphere—acceptance of religious diversity, pluralism and ecumenicity, while at the same time being eclectic and implicit? This seems indeed to be the case, as an inclusive and expansive approach to urban religiosity that can be qualified as a modern stance, appropriate for the highly individualized and varied religious practices one finds in a metropolitan area. Western urban culture provides open and unrestricted possibilities for new forms of spirituality: from more cognitive-oriented reflective environments for self-actualization through Zen or meditation or mindfulness centers, for example, to more communal, participatory and corporeally related forms of religion in which the senses and emotions are more intensely addressed. An example of such new, open religious expressions is the Sunday Assembly movement. This is a global network of people—"a godless congregation that celebrates life"—that wants "to help everyone find and fulfill their full potential," which has been described as a utopia-like (or heavenly) existence.[39] The creation of "Urban Ashrams" is another example of this phenomenon, seeking to bring in the Hindu notion of a spiritual meditative site, a rural refuge, into Western urbanity (cf. Bivins 2015: 142–5). In 2011, this spiritual concept became explicitly connected to the Coltrane movement with the establishment of "The Coltrane Memorial Urban Ashram," a vehicle to achieve *A Love Supreme* life-style: "through devoted practice of universal truths we are living examples of the transformative power of the teachings of A Love Supreme."[40] It is in a similar, open and individualized way, accommodating the often holistic world views of modern urban residents, that Coltrane devotees from all over the world can relate in their own way to the weekly services on Fillmore Street.

Although the St. John Coltrane movement is often depicted by journalists in sensationalistic and superficial ways—as a fringe group of Coltrane zealots or

Figure 10.3 Jamming and meditating on the music of *A Love Supreme* at the John Coltrane Church, August 2011. Photo by Peter Jan Margry.

an eccentric jazz version of Elvis devotion—its veneration of Coltrane's life and music in fact mirrors the attitudes among jazz devotees and others about Coltrane as both a cultural and spiritual icon and inspirational source. For many, Coltrane represents one of the great individual achievements in American music and African American culture, a symbol of creative exploration, self-expression and excellence achieved through discipline, inspiration and hard work. For others, Coltrane embodies the freedom sought for by the Civil Rights movement and black cultural liberation—an example of African American equality and triumph in the face of racism and oppression—just as jazz as an urban phenomenon offered moments of equality and transcendence. The ethos of the church and the religious dimensions of the jam sessions also reflect Coltrane's own deep spirituality and his perception of the transcendent and spiritual dimensions of music and musical performance. Thus, for many visitors to the church, the incarnation of the divine in John Coltrane and his music is present, regardless of whether such a belief is actually couched in other terms due to the current affiliation of the movement with the African Orthodox Church, and is expressed as well by the Coltrane Church itself: "We thank God for the anointed universal sound that leaped (leapt) down from the throne of heaven out of the very mind of God and incarnated in one Sri Rama Ohnedaruth the mighty mystic known as Saint John Will-I-Am Coltrane."[41]

Jazz enthusiast Gary T. (age 38) from Chicago described his experience of the divine in Coltrane's music in the following way:

> I'm not particularly religious, but I heard about the Coltrane Church, so checked it out. It sounded like a wild-ass combo of soulful jazz and serious spirituality. And it was, a righteous scene, and also a real welcoming vibe, people dancing and jamming. The sound just swallows you and the preaching connects. Those preacher-players are deep six in soul. The music gave me a taste of God, Coltrane's sound as a sacred thing.[42]

Eric Williamson (age 43), from Charlottesville, Virginia, no longer attends church services or practices religion within an institutionalized context, but says that his visit to the Coltrane Church evoked a sense of spiritual euphoria and shared joy:

> Coltrane's music in particular is uplifting simply because of its genius. I personally think to celebrate the best and most exceptional in humanity is to celebrate the Creator . . . The experience was genuine and appealed to a sense of the sublime. I felt a sense of being united with others in the abstract and being a part of something bigger for a brief time. It definitely felt therapeutic.[43]

In a similar manner, jazz commentator Gordon Polatnick (2000) described his experience of Coltrane's music as one of belonging, divine love, and faith:

> Sometimes I think I'm the only one who understands what true religion is. It's that cozy state of mind where nothing is more apparent than the unassailable fact that each of us belongs here on Earth, and is deeply loved by an enduring spirit. If you've got that kind of religion, you want to share it.

Conclusion

As no other jazz musician, John Coltrane symbolizes the consanguinity and fusion of jazz and religion, a symbolic interrelationship that is still mediated in the present. Although the cultus of Coltrane is formally embedded within the Christian framework of the African Orthodox Church, research on the weekly practices of the Coltrane movement shows that the musical services devoted to him have a much wider character related to multi-faith and secularist metropolitan urbanicity. The urban renewals of the old Western Addition and Mission districts of San Francisco and the church's forced relocation to a new part of town in the year 2000 weakened its connection to the mainly African American community it had served for years and jeopardized its future, with a decline of locals as well as travelers at the services. Considering the obstacles that the church has faced over the decades—as a jazz-based, non-mainstream African American religious movement that has survived gentrification, exorbitant rents, "cult" accusations, the loss and alienation of community—it is remarkable that the Coltrane Church still exists at all after 45 years. Indeed, its future remains uncertain, as the ongoing gentrification process has pushed out much of the older community and the church continues to struggle with financial issues. However, since the church's return to the heart of the Fillmore district in 2007, weekly attendance seems to be on the upswing, with a flow of visiting jazz enthusiasts and travelers that have added an increasingly dominant global dimension to the services, and who also drop a few bucks into the church donation basket which is circulated throughout the service.

Nonetheless, those who attend the religious performances at St. John's—regular participants and many visitors alike—do not just worship the Christian God, or in fact may not even do that at all: they celebrate the divine genius of John Coltrane and his jazz music as a religious expression. Participants regularly explain how the worship services are transformative, a communion with jazz as sacred music, in a personalized and welcoming communal context. As a grassroots expression of contemporary spirituality, the individualized and inclusive aspects of the Coltrane movement provide insights into the nature of urban religious expression in the twenty-first century. The eclectic and hybrid spirit of the Coltrane Church and the religiosity evoked by Coltrane's music are necessarily products of the urban milieu, where musical forms like jazz and an open orientation towards alternative and multiple religious expressions have been mutually productive and synergetic. Although limited in size, the movement's "churched" status, recently situated in a renovated, gentrified and cool San Francisco district now attracts new audiences open for spiritual experiences and thus allows the movement to survive and spread Coltrane's message and musical charisma worldwide. In this context, the church has created an expansive religious podium where a great variety of devotees and spiritual travelers can find comfort and meaning within the urban environment through the collective celebration of jazz as the divine expression of John Coltrane.

Notes

1 Oriah Vaughn, personal communication to authors, June 6, 2015.
2 Although the religious origins and dimensions of jazz are most frequently associated with Christianity, musicians have drawn upon numerous other religious traditions as well, including Islam, Judaism, Bahá'í, and Buddhism (see Bivins 2015; Stowe 2010).
3 For a discussion of jazz communitarianism, see Bivins (2015: 112–47).
4 See, among others, Peretti (1997); Floyd (1996); Gennari (2006); DeVeaux (1997); Murray (1989); Porter (2002); Neal (1999: 1–23); Ramsey (2003); Berliner (1994); Monson (1996); Levine (1977: 155–89, 292–7); Rosenthal (1992); Saul (2003); Stowe (2010), and Fischlin et al. (2013).
5 Cf. Leonard (1987); Stowe (2010: 312–14) also uses metaphors in discussing the relationship between jazz and religion, not making clear to what extent jazz could be religion in itself and not only an avenue by which existing religious forms can be mediated.
6 During the production process of this volume, the Coltrane Church was forced to move its location once again due to further gentrification and an increase in rent. Eventually, around April 2016, the church found a new shared space a few blocks

away at 2097 Turk Street (Western Addition), in the St. Cyprian's Episcopal Church and its arts community center.

7 Franzo Wayne King was raised in Los Angeles and comes from a family of preachers (the Church of God in Christ, an African American Pentecostal denomination), reared in the tradition of the Pentecostal school of black homiletics. Initially, he worked as a hairdresser, and then after establishing the Coltrane movement, he studied with the leaders of the African Orthodox Church in Chicago in the early 1980s, receiving his Doctor of Divinity degree in 1984. King is an experienced saxophonist as well.

8 Before the church found its most recent storefront housing on Fillmore Street, it used an upper room in the St. Paulus Lutheran Church at 930 Gough Street and operated a co-ministry, between 2000 and 2007. For a history of the church up to the year 2000 as recorded by one of its members, see Baham (2001); his revised and updated analysis is the first book-length study of the church and its role as a community institution (Baham 2015).

9 See US Census Bureau at http://quickfacts.census.gov/qfd/states/06/06075.html; accessed August 15, 2016.

10 By 2013, the median price of a home in San Francisco had reached $1,000,000. See Elise Hu (2013), How This Bay Area Tech Boom's Different From the Last One. *All Tech Considered* (National Public Radio), 13 December; http://www.npr.org/sections/alltechconsidered/2013/12/17/251996835/how-this-bay-area-tech-booms-different-from-the-the-last-one; accessed 12 March 2015; also Joe Kloc (2014), Tech Boom Forces a Ruthless Gentrification in San Francisco, *Newsweek* (April 15); http://www.newsweek.com/2014/04/25/tech-boom-forces-ruthless-gentrification-san-francisco-248135.html; accessed March 12, 2015.

11 Hanna Morjan, personal communication to authors, May 2, 2015.

12 www.westbaysf.org/about.htm; accessed April 6, 2014.

13 Liner notes to *A Love Supreme* (1965).

14 In the late 1960s, prior to the foundation of the One Mind Temple Evolutionary Transitional Body of Christ, the group was first called the Yardbird Club and then the Yardbird Temple, beginning as a small "listening clinic," dedicated to saxophonist Charlie "Yardbird" Parker and John Coltrane; the name was then changed to the Vanguard Revolutionary Church of the Hour in 1969, identifying as a black nationalist group with paramilitary leanings aligned with the Black Panthers, emphasizing social change and revolution, and influenced by the ideas of Dr. Huey Newton according to founder Franzo King, who explained that the church eventually shifted its emphasis from revolution to "spiritual evolution" (Washington 2001: 408–9).

15 See Freedman (2007); Washington (2001: 394); Saint John Coltrane African Orthodox Church website: http://www.coltranechurch.org/#!about/csgz; accessed April 3, 2014.

16 G. Duncan Hinkson left the African Orthodox Church in 1984 and created his own independent jurisdiction, the African Orthodox Church of the West, consecrating Franzo King as a bishop.

17 https://m.facebook.com/stjohncoltranechurchwest/about?expand_all=1; accessed April 1, 2006.

18 King quoted in the documentary *The Church of Saint John Coltrane,* by Gayle Gilma and Jeff Swimmer (Tango Films, 1996).

19 For the mission of the Coltrane Church, see its Facebook page at: https://m.facebook.com/stjohncoltranechurchwest/about?expand_all=1; accessed August 15, 2016.

20 Oriah Vaughn, personal communication to authors, June 6, 2015.

21 Implicit religion is a concept proposed by sociologist Edward Bailey: an expression of the *seemingly* secular (in this case, jazz music) that contains a religious dimension of its own (Bailey 1997; cf. Margry 2012; Bivins 2015).

22 See the interview with Dukes and Franzo King in 2010 at http://openspace.sfmoma.org/2010/01/interview-with-coltrane-icon-painter-mark-dukes-and-archbishop-franzo-king/; accessed August 15, 2016.

23 Lyrics from "Lush Life," performed on the 1963 studio album *John Coltrane and Johnny Hartman* (Strayhorn 1997).
24 Prior to the noon worship services, Deacon Marlee-I Mystic leads "A Love Supreme Meditation" at 10:30 a.m. that involves the recitation of Coltrane's spiritual testimony from his liner notes of *A Love Supreme* and concludes with the singing of the devotional poem he wrote, the words of which correspond to the music of the fourth movement, "Psalm." Apart from the weekly services, the Reverend Wanika King-Stephens and Marlee-I Mystic also host the "Saint John Will-I-Am Coltrane Uplift Broadcast," which airs every Tuesday from noon to 4 p.m. on KPOO 89.5 FM. The show includes the music and inspirational words of wisdom from John Coltrane.
25 The fieldwork for this essay was executed in the summers of 2011 and 2012, fall 2012, summer 2013 and spring 2015.
26 This happens regularly during the winter when there are fewer people visiting San Francisco and the church.
27 See the Saint John Coltrane African Orthodox Church website: http://www.coltranechurch.org/#!services/c8k2; accessed August 15, 2016; on the polymorphous sound of jazz as prayer, see Bivins (2014).
28 The African Orthodox Church celebrates the seven sacraments of the Roman Catholic Church, follows Eastern and Western rites and liturgies, and affirms the Nicene Creed, the Apostles' Creed, and the Athanasian Creed in an amalgamation of various religious traditions.
29 http://www.yelp.com/biz/the-st-john-coltrane-african-orthodox-church-san-francisco; accessed April 6, 2014.
30 The African Orthodox Church began as a denomination for black Episcopalians, founded in 1921 by George Alexander McGuire at time when people of color were excluded from advancement in the Episcopal Church. Its early membership was primarily West Indian; the AOC currently has approximately 2,500 members organized around 17 parishes throughout the world, in which the St. John Coltrane Church is the only branch in the western United States (Pruter 2006: 85). As an independent movement that arose in response to the exclusionary racism of white churches at the time, the AOC has a history of opposition and controversy, as its founder McGuire was consecrated a bishop in 1921 by the controversial renegade *episcopus vagans* ("stray bishop") Joseph René Vilatte (1854–1929), a consecration enacted to place McGuire and the AOC in apostolic succession, but an apostolic status condemned as schismatic by the Syriac Orthodox Patriarchate of Antioch and All the East (Brandreth 1987 [1947]: 70).
31 http://www.coltranechurch.org/#!mission/c21kz; accessed February 20, 2015.
32 http://www.yelp.com/biz/the-st-john-coltrane-african-orthodox-church-san-francisco; accessed April 6, 2014.
33 King quoted in Boulware (2000); accessed April 6, 2014.
34 http://www.coltranechurch.org/#!about/csgz; accessed February 20, 2015.
35 Quotations from the documentary *The Church of Saint John Coltrane* by Gayle Gilma and Jeff Swimmer (Tango Films, 1996).
36 King quoted in Cox (1995: 154).
37 Text on backside of Coltrane's album *Meditations* (1966), the spiritual follow-up to *A Love Supreme*.
38 Marlee-I Mystic, personal communication to authors, April 11, 2014.
39 See, for example: http://soulpancake.com/conversations/view/6789/does-religion-enable-humans-to-reach-their-full-potential.html; accessed August 15, 2016 and https://en.wikipedia.org/wiki/Scientology_beliefs_and_practices; accessed August 15, 2016.
40 https://www.facebook.com/pages/Coltrane-Memorial-Urban-Ashram/328929710491994; accessed August 15, 2016.
41 At the beginning of the Coltrane movement, it was believed that Coltrane was a god or an incarnation of God, but as of 1982 Coltrane became Saint John within the AOC and

since then his music is perceived a divine incarnation; cf. http://www.coltranechurch.org/#!about/csgz; accessed August 15, 2016.

42 Gary T., personal communication to authors, April 25, 2015.

43 Eric Williamson, personal communication to authors, May 31, 2015.

References

Baham, Nicholas Louis, III (2001), Out of this World. Anthropological Testimonies of Awakening and Renewal in Coltrane Consciousness. An Ethnography of the St. John Will-I-Am Coltrane African Orthodox Church, PhD Dissertation, Indiana University.

Baham, Nicholas Louis, III (2015), *The Coltrane Church: Apostles of Sound, Agents of Social Justice*. Jefferson, NC: McFarland.

Bailey, Edward I. (1997), *Implicit Religion in Contemporary Society*. Kampen, Netherlands: Kok Pharos.

Berkman, Franya J. (2007), Appropriating Universality: The Coltranes and 1960s Spirituality, *American Studies*, 48(1): 41–62.

Berliner, Paul (1994), *Thinking in Jazz: The Infinite Art of Improvisation*. Chicago, IL: University of Chicago Press.

Bivins, Jason C. (2014), Take It to the Bridge: Jazz Prayers, *Reverberations: New Directions in the Study of Prayer* (March 28); at SSRC Social Science Research Council Forums; http://forums.ssrc.org/ndsp/2014/03/28/take-it-to-the-bridge-jazz-prayers/; accessed August 15, 2016.

Bivins, Jason C. (2015), *Spirits Rejoice! Jazz and American Religion*. New York: Oxford University Press.

Boulware, Jack (2000), Requiem for Church Supreme, *SF Weekly* (January 26); http://m.sfweekly.com/sanfrancisco/requiem-for-a-church-supreme/Content?oid=2137874; accessed April 6, 2014.

Brandreth, Henry R. T. (1987) [1947], *Episcopi vagantes and the Anglican Church*. San Bernardino, CA: Borgo Press.

Brown, Leonard L. (ed.) (2010), *John Coltrane and Black America's Quest for Freedom: Spirituality and the Music*. Oxford: Oxford University Press.

Cox, Harvey (1993), Jazz and Pentecostalism, *Archives de sciences sociales des religions*, 38: 181–8.

Cox, Harvey (1995), *Fire From Heaven: The Rise of Pentecostal Spirituality and the Reshaping of Religion in the Twenty-First Century*. Reading, MA: Addison-Wesley.

Davis, Francis (2001), Coltrane at 75: The Man and the Myths, *New York Times* (September 23), Arts Section: 25, http://www.nytimes.com/2001/09/23/arts/music-coltrane-at-75-the-man-and-the-myths.html?pagewanted=all; accessed August 15, 2016.

DeVeaux, Scott Knowles (1997), *The Birth of Bebop: A Social and Musical History*. Berkeley: University of California Press.

Fischlin, Daniel, Ajay Heble and George Lipsitz (2013), *The Fierce Urgency of Now: Improvisation, Rights, and the Ethics of Cocreation*. Durham, NC: Duke University Press.

Floyd, Samuel A. (1996), *The Power of Black Music: Interpreting Its History from Africa to the United States*. New York: Oxford University Press.

Freedman, Samuel G. (2007), Sunday Religion, Inspired by Saturday Nights, *New York Times*. December 1; http://www.nytimes.com/2007/12/01/us/01religion.html?_r=0; accessed August 15, 2016.

Gennari, John (2006), Jazz in African-American Culture, in Colin A. Palmer (ed.), *Encyclopedia of African-American Culture and History*, 2nd edn. Detroit, MI: Macmillan Reference USA, Vol. 3, 1167–9.

Heelas, Paul and Linda Woodhead (2005), *The Spiritual Revolution: Why Religion is Giving Way to Spirituality*. Oxford: Blackwell.

Klein, Jordan (2008), A Community Lost: Urban Renewal and Displacement in San Francisco's Western Addition District, MA Thesis, University of California, Berkeley.

Leonard, Neil (1987), *Jazz: Myth and Religion*. New York: Oxford University Press.

Levine, Lawrence W. (1977), *Black Culture and Black Consciousness: Afro-American Folk Thought from Slavery to Freedom*. New York: Oxford University Press.

MacDonald, Eli (2015), Taking the Pulse of the Fillmore, *San Francisco Foghorn* (February 11); http://sffoghorn.org/2015/02/11/taking-the-pulse-of-the-fillmore; accessed August 15, 2016.

Margry, Peter Jan (2012), European Religious Fragmentation and the Rise of Civil Religion, in Ullrich Kockel, Jonas Frykman and Máiréad Nic Craith (eds), *A Companion to the Anthropology of Europe*. Malden and Oxford: Blackwell Wiley.

Monson, Ingrid (1996), *Saying Something: Jazz Improvisation and Interaction*. Chicago, IL: University of Chicago Press.

Murray, Albert (1989), *Stomping the Blues*. New York: Da Capo.

Neal, Anthony Mark (1999), *What the Music Said: Black Popular Music and Black Culture*. New York: Routledge.

Nisenson, Eric (1995), *Ascension: John Coltrane and His Quest*. New York: Da Capo.

Peretti, Burton W. (1992), *The Creation of Jazz: Music, Race, and Culture in Urban America*. Urbana: University of Illinois Press.

Peretti, Burton W. (1997), *Jazz in American Culture*. American Ways Series. Chicago, IL: Ivan R. Dee.

Polatnick, Gordon (2000), The Jazz Church; http://elvispelvis.com/jazzchurch.htm; accessed August 15, 2016.

Porter, Eric (2002), *What Is This Thing Called Jazz? African American Musicians as Artists, Critics, and Activists*. Berkeley, CA: University of California Press.

Porter, Lewis (1998), *John Coltrane: His Life and Music*. Ann Arbor: University of Michigan Press.

Pruter, Karl (2006), The African Orthodox Church, in *The Old Catholic Church*. Rockville, MD: Wildside Press.

Ramsey, Guthrie P., Jr. (2003), *Race Music: Black Cultures from Bebop to Hip-Hop*. Berkeley: University of California Press.

Rosenthal, David (1992), *Hard Bop: Jazz and Black Music, 1955–1965*. New York: Oxford University Press.

Santoro, Gene (1992), Chasin' de Trane, *The Nation*, April 13, 497–9.

Saul, Scott (2003), *Freedom Is, Freedom Ain't: Jazz and the Making of the Sixties*. Cambridge, MA: Harvard University Press.

Solnit, Rebecca (2000), *Hollow City: The Siege of San Francisco and the Crisis of American Urbanism*. London and New York: Verso.

Stark, Rodney and William Sims Bainbridge (1981), Secularisation and Cult Formation in the Jazz Age, *Journal for the Scientific Study of Religion*, 20(4): 360–73.

Stowe, David W. (2010), Both American and Global: Jazz and World Religions in the United States, *Religion Compass*, 4(5): 312–23.

Sylvan, Robin (2002), *Traces of the Spirit: The Religious Dimensions of Popular Music*. New York: New York University Press.

Washington, Michael Spence (2001), Beautiful Nightmare: Coltrane, Jazz, and American Culture, PhD Dissertation, Harvard University.

Whyton, Tony (2013), *Beyond A Love Supreme: John Coltrane and the Legacy of an Album*. Oxford: Oxford University Press.

Woideck, Carl (1998), *The John Coltrane Companion: Five Decades of Commentary*. New York: Schirmer Books.

11 "Singing is prayer two times"

A transnational perspective on "religious music," musical performance and urban religiosity in Cameroon and Switzerland

Raphaela von Weichs

Music is a cultural product and has followed routes of migration (Connell and Gibson 2003: 160), entering through urban centers into rural areas, but also taking the reverse path. Like music, religion has a mobile and transformative dimension (Hüwelmeier and Krause 2010; Rytter and Olwig 2011; Shankar and Adogame 2012) and brings new dynamics to urban contexts, especially when combined with new media. As a cultural product, music has been used as a medium for Christian proselytization, and in recent years "religious music" has intensified as a means of evangelization emanating from urban centers. This is noticeable in the many ways religion meets new media (Campbell 2010), in the vibrant marketing of "religious" (read "Christian") media productions and their local and global dissemination, and in the growing popularity of gospel groups and choir associations as a global movement. However, music may not only be a means for religious conversion, it can also be a form of prayer (for example, Berliner 1975/76: 132; Rosenwein 2000: 42), and a form of ritual communication with spiritual beings (Merriam 1964: 217). In this chapter, I will focus on Christian choral music as a form of prayer and place-making in the migration of Cameroonians to Europe. I do this by analyzing the interplay between processes of religious innovation and spiritualization in urban Africa and the invention of religious events for the migrants within semi-rural and urban areas of Europe.[1]

Directing attention to the interplay of music and religion in transnational migration allows us to analyze the symbolic and cultural capital that is mobilized in situations of transition, and the transformative aspects of music in urban and religious places. Music as well as religious events provide stages for cultural representation. Through participation in such events, the involved actors learn and share new cultural practices. Though cities have been identified as places of intense social and cultural dynamics (Pinxten and Dikomitis 2009; Livezey 2000, for example), I argue that religious innovation and transformation is not confined to the city. It also results from a constant interplay between urban and (semi-) rural places. Processes of urbanization and migration have dissolved the sharp divide between cities and rural areas, as Hegner and Margry state in the Introduction to this volume. This vanishing of contrasting lifestyles and habitats has generated areas of the "urbanesque." In Europe, regions like the *arc lémanique* in Switzerland or the *Ruhrgebiet* in Germany have developed into zones where non-migrants and migrants share social, cultural, economic and political settings. The urbanesque becomes an ever-growing space of cultural encounters, which shapes our understanding of the urban and the rural. In this space, cultural and religious events take place in order to express and negotiate cultural diversity in multiple ways (Salzbrunn: 2014).

This chapter builds on the pioneering work of anthropologists, sociologists and geographers who studied the transformative force of religion in urban places and urban spaces. Religious studies scholars have adopted a global perspective to conceptualize an emerging global denominationalism that has generated new forms of hybrid religions, such as Bahá'í, Moonies, Hare Krishnas, African American religions, and that transforms modern urban life (Becci et al. 2013: 15). This process can be observed in Switzerland, where 21.4 percent of the total population were non-confessional in 2012 (as opposed to 7.4 percent in 1990) and 12.2 percent belonged to other than Christian religious communities (as opposed to 3.7 percent in 1990) (SPI St. Gallen 2014; Swiss Federal Statistical Office 2003). The Catholic Church shrunk from 46.2 percent of the total population in 1990 to 38.2 percent in 2012. Figures vary widely between the cantons and cities, but urban Catholic parishes in particular are losing members. The Swiss Catholic Church aims to make up the loss through the incorporation of new migrants from Africa, Asia and America. Stokes (1994: 11) argues that music and performance help to shape the related processes of place-making and identity creation. He also considers music as a field of symbolic activity that is highly important to state authorities (Stokes 1994: 10–11). Such reasoning can be extended to churches which need to compete with the symbolic productions of rivaling religious actors and movements, and need to (re-)invent forms of incorporation. Pilgrimages and (choir) music form important symbolic activities of Catholic mission and parish work. This is especially so in urban places where new migrants search for entry and contact points with locals and local cultures. Connell and Gibson's more recent critical geographies follow soundtracks of global and transnational music styles to identify their impact on popular culture, identity, consumerism, and place-making (Connell and Gibson 2003: 4). Yet, religion and religious music are not part of their analysis. Most urban studies dealing with religion and music focus on the transformation of the city through either music *or* religion. Few, however, look at the intricate relationship between music *and* religion or religiosity in the city. They also neglect the translocal and transnational dimension of music and religion that transforms urban and religious spaces. In this regard, we need to take account of social processes that link the urban and the rural as well as the space in between: the urbanesque habitat.

I will now first discuss "religious music" as a genre which forms and transforms translocal urban spaces to see how music has taken form in different cultural, religious and urban contexts. I will then look at the choir landscape in Yaoundé, the capital of Cameroon, to show how religious music becomes a matter of cultural practices, politics and evangelization both in, and emanating from, urban centers. My particular focus is on singing, considered "as prayer two times" in Cameroon. Finally, I will link Cameroonian religious and choral music with the invention of

"African pilgrimages" in Switzerland. There, I am looking at the role of music and performance in transforming urban religiosity and cultural practices from a transnational perspective.

Theorizing and contextualizing "religious music"

The term "music" is as broad and vague as the concept of "religion." Further, the two do not lead to more clarity when combined. However, "religious music" has evolved as a generic and emic term for a certain type of music—namely, music that is based on sacred texts such as the Bible or carries religious meanings, and is used in religious rituals or that transmits religious messages and spiritual devotion. The label "religious music" appears on DVDs and CDs that can be bought on the sidewalk or in music stores in urban Cameroon. This label indicates the moral value that Christians give to religious music. Stokes, who focuses on the musical construction of place, argues that "music is socially meaningful, not entirely but largely because it provides means by which people recognize identities and places, and the boundaries which separate them" (Stokes 1994: 5). He defies any essentialist definition of music and argues that "music 'is' what any social group considers it to be" (Stokes 1994: 5). His constructivist approach is shared by migration scholars, who debate the contributive and transformative dimension of music that is brought and performed by immigrants in the first and subsequent generations. Music is not only produced for listening and dancing, but may in itself be a form of political action (Martinello and Lafleur 2008: 1194).

In this chapter, I take religious music as a cultural performance that is produced by immigrants in urban habitats and through performativity (Fischer-Lichte 2012). This relates to the way it is performed, and produces and transforms social reality, meaning and sentiments in religious contexts. According to the *Oxford Music Online Encyclopaedia*, "music in many religious traditions is considered to have sacred qualities and 'sacred music' is taken to mean the musical settings of sacred texts." However, "there is a deep theoretical and theological tension" within the Catholic Church concerning the role of music in the liturgy (Marini 2012). This tension dates back to the time of Bishop Augustine of Hippo, the early Christian theologian and philosopher. For Augustine, music carries the deepest emotions possible in the adoration of God, but it also focuses attention on the individual's sensations, and needs to be disciplined (Ostrem and Petersen 2013: 1433). This implies a disciplining and control of the body. Catholic sacred music therefore contrasts with music that induces ecstasy, possession, or trance, as is the case in many African religious traditions. "Ever since Augustine," Marini continues, "Catholic tradition has warned against artistic excess in worship music" (Marini 2012). As we will see later on, this tension is reproduced in the cultural and religious encounters of African Christians who actively perform music in the Swiss Catholic Church. The incorporation of African choir performance into Swiss Roman Catholic liturgy at St. Maurice, an Augustinian monastery, was linked to the question of how this music should be performed to stimulate sensation and a devotional mood. St. Augustine is popularly, especially in Catholic circles, credited with the words "he who sings prays twice,"[2] although this quote is not literally attested in Augustine's own texts and sermons.

In Cameroon, music is valued for the sensation it gives to the whole body. Body movement is part of worship through singing. The expression "singing is prayer two (or many) times" is omnipresent among Christians. In this chapter, "religious music" is understood to be a mode of normative communication and of expressing religious feelings, based on spirituality or religiosity. Basically, "religious music" in Cameroon is an urban invention. It is a label given to music that is considered Christian, civilized and evangelical. It is also supported by digital devices and produced in digital studios. It goes beyond European sacred musical traditions. It includes African musical traditions, tunes, rhythms and languages, as long as the message is evangelical. In the following, I focus on the entertaining and evangelizing dimension of religious music in Cameroon and in transnational migration from Cameroon to Switzerland. The boom in new religious movements that go beyond Christianity and of new churches that innovate and reinvent Christianity is visible in Africa and Europe (Adogame 2013). With new technologies in Africa, such as computer-based music and video production, it has generated a new genre and a market for "religious music" as part of popular culture. While collective singing, dancing and music traditions, of course, existed in Africa long before colonization, and are therefore nothing "new," modern choral music only emerged at the end of the nineteenth century (de Beer and Shitandi 2014: 185). It was born out of the merging of African, European and American musical traditions, particularly through colonial and missionary education. It became more innovative and popular in the 1950s and has boomed since the 1990s with the rise of evangelical Christianity (de Beer and Shitandi 2014: 186, 198). Choral music in Cameroon is strongly influenced by European and American traditions. It is now a widespread practice not only in cities such as Yaoundé, Douala, or Buea, but also in semi-urban places like Limbé and Kumba. These urban and semi-urban places were, and are, centers of mission stations and missionary activities.

Furthermore, a number of urban and anthropological studies has shown that cities are gateways for migrants, providing many sites and opportunities for gatherings and cultural exchange. Glick Schiller has argued that not only mega-cities but also large, medium and smaller cities provide locally specific opportunity structures for migrant incorporation. Cultural and religious groups and institutions like churches, choirs and music bands may facilitate the place-making of newcomers. They may also function as incorporating structures (Glick Schiller 2009: 130). Further, cities have been identified as nodal points of exchange and innovation in music (Kiwan and Meinhof 2011: 5). Choral music as a genre of religious music comes from different musical and cultural traditions. Additionally, it is performed in both religious and non-religious contexts. It continues to occupy an significant place in churches and the entertainment industry in Africa today (de Beer and Shitandi 2014: 193). It is also of growing importance in cultural practices and in life-cycle rituals, like funerals and weddings. In this regard, it has supplemented and partly replaced former musical traditions and musical practices (Jindra 2011). As a transcultural practice that transgresses national, social, ethnic and religious boundaries, it expresses the experience of Christian spirituality. It is transferred onto migration, incorporated and transformed or adapted to the social and religious settings. The new popularity of "religious music" in Cameroon corresponds to processes of religious restructuring in Christianity. It addresses religious, social and cultural aspects of community life. Therefore, it is not unusual to find religious choir groups and associations in big cities who go to urbanesque and rural sites to evangelize.

The production, performance, politics and diversity of "religious music" in urban Cameroon

I will now look at the religious music landscape as I encountered it in Cameroon in 2012. During four weeks of ethnographic fieldwork in the cities Yaoundé, Duala and Buea, I conducted research on choirs of the Catholic, Presbyterian, Evangelical and Baptist Churches, as well as on non-denominational university choirs.[3] The production, performance and politics of urban choirs in Yaoundé will be analyzed by focusing particularly on two examples: one from the Catholic and one from the Presbyterian Church. I will then discuss how singing seen "as prayer two times" in Cameroon becomes a matter of cultural practices, politics and evangelization emanating from urban centers. In other words, I will examine how music is spiritualizing the city and its environment in this particular cultural and political context.

In Yaoundé, the capital city of Cameroon, Christian gospel singers, choirs and bands, numbering into the hundreds, creatively blend local music with Western musical traditions, such as country or gospel. They record it (often in Nigerian studios) and sell the recordings in music shops, on the street and in cross-country and inter-city buses. Religious movements and new churches like the Evangelical "Redeemed Church of God" use translations of the Bible into local Cameroonian languages and interpret these translations to compose new hymns. The initiative to translate the Bible is a project of the Cameroon Association for Bible Translation and Literacy. Biblical narratives, such as the Immaculate Conception of the Virgin Mary or the interference of witches in the social life of humans, are composed into songs. These are then visualized in music dance videos, borrowing Nollywood iconography of the rising Nigerian video and film industry (Krings and Okome 2013). This is done with the explicit objective of evangelizing not only those in the city, but also those left behind in the villages, who may not speak French (the colonial and post-colonial lingua franca of the urban population). Likewise, urban choirs perform in various neighborhoods of the capital city and travel to other cities in Cameroon, to neighboring African states, and to Europe. They also go to rural sites to perform their religious and secular repertoire. The choirs can be hired for personal, political and other purposes, to enliven both public and private events. In this way, religious choirs get out of their institutional environment, and link urban, semi-urban and rural places through various cultural activities.

One outstanding example of the multi-dimensional activity of urban religious choirs is *La voix du cénacle*, a choir founded in the capital city Yaoundé by the former managing director of Cameroon Radio and Television (CRTV), Gervais Mendo Zé. *La voix du cénacle* started as a Catholic prayer group and evolved into a choir with fifty members. It became well-known through its religious and non-religious performances in the CRTV program. Professor Mendo Zé, educated at a Catholic missionary school, became a respected linguist at Yaoundé University, and acquired public recognition as a composer of hymns and a playwright.

Belonging to the Catholic Church and to the Beti ethnic group, like President Paul Biya, Mendo Zé shared in the political power and privileges of the Beti/Boulu as long as he served as a manager of CRTV.[4] Under his management, CRTV turned into a voice of the ruling party. Mendo Zé was deeply ingrained in the Catholic Church, the strongest of the former mission churches. Thus, the performance of his choir in the central national media also signaled Catholic hegemony in a heterogeneous and competitive field of urban and rural religiosity. *La voix du cénacle* was invited by the political elites of Cameroon and by neighboring African countries to perform at social, political and religious events, such as weddings, funerals and the inauguration of religious and political buildings. On the international scene, the choir won several prizes in music competitions, such as the "Golden Palme of the International Festival of Sacred Music" in Rome in 2005.[5] By composing praise hymns like "Le choix du peuple" and "Chantal Biya," Mendo Zé supported the President's campaigns and the charitable activities of the First Lady. His blurring of religious music performance and political propaganda, all broadcast in public, gave President Biya and his regime an almost religious blessing.

Like other successful choirs, *La voix du cénacle* produced and marketed digital music recordings and videos mainly through urban, translocal and transnational networks. Talented, entrepreneurial figures, such as Professor Mendo Zé, are founders and managers of successful choirs. In the mainly urban setting of his choir's activities, he mobilized financial resources and political and religious networks to promote his reputation as a composer and as the manager of one of the most famous choirs in Cameroon. Choir leaders in Cameroon constantly need to find funding as well as ways to sustain, promote and commercialize their choir's performances. To achieve these ends, they maintain networks in the political, economic and religious circles both in the country and in the diaspora. They compete in music competitions to enhance their choir's reputation and to win awards. Such competitive festivals are organized on multiple levels by religious and non-religious institutions. They also occupy an important place in educational institutions like schools and universities.

In Yaoundé, choirs may be organized on an ethnic or geographic basis, with a majority of members sharing a common cultural or regional background. These may people who are already residents of the city or they may be newcomers who contribute local songs, sounds, rhythms and instruments as cultural capital to the group. In the Presbyterian Church in Yaoundé's inner-city quarter Nsimeyong, my second example, I found a number of choirs holding rehearsals at the same time on different levels in and outside of the church building. These choirs were organized on the basis of age, gender, cultural, ethnic identity and urban residency.

The "Holy Trinity Choir," with a mixed gender and ethnic membership, practiced gospels, spirituals and hymns on the ground floor. Next door, the "Praise and Worship Band" was accompanying vocalists with electric guitars, piano and percussion. On the upper floor, I found the women's fellowship choir, with various ethnic memberships, next to the "Hallelujah Choir" made up of north-westerners, men and women of various ages who also used their rehearsals as religious and cultural meetings. All in all, this church had ten different choirs: five organized along ethnic and linguistic lines, four according to age and gender (the fellowships of men, women and youth), and one large gospel choir with a mixed membership. With this diversity in music, performance and membership, choir members gain access to urban networks, and are supported in their construction of urban identities. For some of them, the idea of using musical performances as a way of urban and religious incorporation is transferred onto transnational migration. I will come back to this aspect in the second part of my chapter.

For their performances, choirs have a repertoire consisting not only of music but also of costumes symbolizing local, national, social, or other collective and religious identities. Female choristers wear the *Kaba*, a dress which covers the whole body.[6] These costumes are usually worn with headdresses and may have imprints, patterns, or religious emblems like the Virgin Mary on them, symbolizing social and religious belonging. Once a dress for festive and ritual events of the Sawa in Southern Cameroon, the *Kaba* was considered a chaste garment for native Christian women by European missionaries. Though scorned in the late colonial days by Cameroon's younger women, the *Kaba* experienced its comeback in the wake of a search for national identity, and as a garment for religious music choirs.

Considering the long Catholic tradition of warning against "artistic excess in worship music" (Marini 2012), body movement in Cameroon is more constrained in the Catholic than in most of the Protestant and new churches. Further, choirs wear either Western-style combinations in two colors or the *Kaba* style, depending on the nature of their performance. Choirs have become part of religious rituals, like funerals and weddings, not only in the cities, but also in the villages. As Jindra reminds us, Christian choir participation has become such a standard practice of "death celebrations" in the Cameroonian Grassfields that these choirs have almost replaced the traditional dance societies (Jindra 2011: 109). They have even become a financial burden for the inviting party, and a sign of the social and political transformation in the villages. "Death celebrations" are no longer held exclusively for deceased dignitaries, the prospective ancestors, but organized for anybody who can afford it. As I was told by choir leaders from the Catholic and Presbyterian Churches and the "Redeemed Church of God," a Pentecostal Church branch from Nigeria, they travel to rural sites to preach the gospel through music and thereby evangelize.

These ethnographic observations show that choirs are a vital part of urban and rural Christian religiosity in Cameroon. They can be found in all types of churches, from established denominations to newly founded ones, restructuring immigrant and resident populations and their affiliations. Under colonial rule,

African music (practices, performances and instruments) had been "demonized and belittled" by European missionaries and white clerics "due to its ties to indigenous 'pagan' traditions and cultures" (Falola and Fleming 2012: 3). This attitude officially changed in the 1960s, when most colonies gained independence, among them Cameroon in 1960. The Second Vatican Council had voted for a strategy of "enculturation" of indigenous (musical) practices into the liturgy.[7] However, this applied only to those and to such a degree that they were compliant with the Catholic doctrine. This transformation of religious practices in the Catholic and other former mission churches gained momentum in the 1990s. Christian fundamentalist churches, particularly new evangelical churches from Nigeria, established branches in Cameroonian cities. They introduced digital technology in liturgical services and produced "religious music" as a means of proselytization. Now, "religious music" was *en vogue* and recorded *en masse*, produced in studios, and sold in the street market and in music shops. More and more urban choirs came up to sing hymns in vernacular and to use traditional instruments, such as the balafon (a xylophone), drums and flutes, with the double aim of spreading the gospel and incorporating newcomers. With this shift to indigenous music cultures, to the composition of Christian hymns in vernacular, and to socio-cultural activities organized by the choir groups, they more easily incorporate rural migrants into the urban congregations. At the same time, the musical repertoire of urban choirs is inspired and renewed by migrants that come to the city. It is then transformed into a Christian version and brought back by the urban choirs to be performed in life-cycle rituals and other social, religious or political events in the rural areas. Therefore, the links between the urban and the rural are never cut. They are in constant flux, reciprocally transforming social structures and bringing new cultural and religious practices to the city and the village. While Orsi (1999) finds a deep-rooted distrust against the sinful city, which simultaneously attracts desires of all sorts in twentieth-century America, in Christian discourse and practice in Cameroon, evil lurks in both the village *and* the city. Television magic and televangelism channels like "Emmanuelle TV" narrate daily that evil is located here and there. Both sites must therefore be expurgated of evil spirits, something that may be achieved through the power of "singing as prayer two times."

Cameroonian "religious music" in African-Swiss pilgrimages

I will now switch my observations to Switzerland and use my ethnographic material from the towns of St. Maurice and Einsiedeln.[8] Here, I will show how similar strategies and techniques of cultural and body performance, similar to those used in Cameroon, are adapted as a way of reverse missionization and place-making in the transnational migration of Cameroonians in Switzerland.

In 2011, inspired by the work of anthropologists on "Travelling Spirits" (Hüwelmeier and Krause 2010) and being a newcomer to Switzerland,[9] I began to study the religious life of transnational migrants. I focused on those in and around the city of Lausanne, the fifth largest city in the Helvetic confederacy (after Zurich, Geneva, Bern and Basel). I learned about two annual pilgrimages organized by the Swiss Catholic Church for migrants from Africa: the *Pèlerinage aux saintes et saints d'Afrique*, which takes place in St. Maurice, a small town in the French-speaking canton Valais, and the *Afrikanische Pilgerfahrt* held in Einsiedeln, a small town in the German-speaking canton Schwyz.

The pilgrimage of St. Maurice was founded and launched for francophone Switzerland in 2002 by a group of Catholic missionaries, the *Comité du groupe de coopération missionaire* (cf. Ballif 2010; Salzbrunn and von Weichs 2013). The *Groupement Coopération Missionaire en Suisse Romande* (GCMSR) is commissioned by the Swiss Bishops' Conference (SBC) to invigorate missionary activities in French-speaking Switzerland.[10] It also has the implicit objective of attracting migrants from Africa (Ballif 2014: 1). The society "Missionaries of Africa," better known as the "White Fathers" because of their white habit, is an active part of the GCMSR and has a branch in the Swiss city of Fribourg. Among other things, the White Fathers take care of migrants from Africa and co-organize the *Pèlerinage aux saintes et saints d'Afrique*. Most of the Swiss White Fathers are now retired, but there are several African White Fathers who rejuvenate the "Missionaries of Africa." Since 2008, the GCMSR has also been supported by *Missio*, a branch of the Catholic Church's official charity for overseas missions.

The pilgrimage is devoted to St. Maurice d'Agaune. The legend surrounding this ancient saint depicts him as an African who is said to have led the Theban Legion of the Roman Empire to Europe. Having refused to kill Christians, he and his followers are said to have been killed around 300 CE by other Roman legionaries (close to the present-day town of St. Maurice) where a monastery and a shrine of St. Maurice were build. The story of his legend is repeated annually at the opening ceremony of the pilgrimage in St. Maurice and used as a device to link the history of Africa and Switzerland. Each year, a different male or female African saint or a group of African martyrs is commemorated at the *Pèlerinage aux saintes et saints d'Afrique*.

As a counterpart to the symbolic importance of martyr narratives and relics at the pilgrimage of St. Maurice, the Virgin Mary takes center stage at the second "African pilgrimage" in Switzerland, in Einsiedeln. This event is organized by *Migratio*, an organ of the Swiss Conference of Catholic Bishops that administers the integration of migrants into the Swiss Catholic Church and their related pastoral care. The Virgin of the Abbey of Einsiedeln is distinct: she has black skin, a golden garment, and her black marble chapel is enclosed by a baroque church. I was told by a Cameroonian pilgrim that the black color of this Virgin's skin was jokingly discussed by other African pilgrims who claimed that it testifies to the "blackness" of God. In the context of a racialized and exoticized image of Africans in Europe, the blackness of God and the Virgin becomes a symbol of empowerment. This contrasts with urban Cameroon, where the "white" Virgin Mary is an important figure of veneration. Here, her whiteness is a symbol of power, though an ambiguous one. "White" refers to the powerful other, the colonialist, and it signifies the powerful self, the ancestors. The veneration of the "white" Virgin Mary is thus both, a relict of European hegemony in the former African colony, and a constitutional part of contemporary African religiosity. As a symbol of power, her whiteness has turned into the opposite for some Cameroonian pilgrims in Einsiedeln.

When I participated in these pilgrimages for the first time in 2011, I noticed that choirs and their performances were at the center of the events. As I have argued previously, religious events organized around musical performances make migrants more visible and make it possible to attract and incorporate them into church structures (Salzbrunn and von Weichs 2013: 8). But why were these events located in places far away from Switzerland's larger cities? And why were they attracting choirs to perform "African chorale music"? Before addressing these questions, I will look at the similarities and differences between the two events.

The two pilgrimages share three common features. They both include, first, a procession (through a residential area of St. Maurice and through the forest path of the stations of the cross in Einsiedeln) and second, a liturgical Mass. Third, they are both guided by music, dance and prayer. Pilgrims travel to these events by bus. Each bus is organized by a choir. Each comes from a different city or canton in Switzerland. Food is prepared and consumed in groups. Though the missionaries in St. Maurice encourage all pilgrims to share a common meal, most choirs eat separately. European pilgrims bring their individual lunch packets. Social and residential belonging is a key marker for identity constructions, since most members live in urban or urbanesque habitats in the Lake Geneva region and some need to travel long distances to reach the pilgrimage sites. As Coleman and Eade (2004: 2) have argued, in many cases, pilgrimages have become cultures in motion rather than rites of passage. As one Nigerian pilgrim told me, his participation in the event was like a trip to "Africa" or "home." This was expressed in the instruments and rhythms of music, the style of food, the collective meal, the chatting, and praying. For him, as an undocumented migrant living in Bex, an urbanesque place in Switzerland, it was also a break from a stressful situation, as well as a space for information exchange and making contacts.

Choirs express their multiple belongings with names like *Chorale Africaine de Fribourg* or refer to their place or region of residence with names like *Chorale St. Joseph de Jura*. Most women are in costume, similar or identical to the Cameroonian *Kaba* style. Men wear wax print suits or shirts. Some choirs print their affiliation on their costume, such as the *Association Camerounaise de paroisse St. Joseph de Lausanne,* thus indicating their urban, transnational belonging. Marian emblems are also powerful symbols of religious belonging, empowerment and affiliation, suggesting that the Virgin Mary provides answers to problems encountered in post-migration situations (Hermkens et al. 2009: 2). Some choirs are heterogeneous, assembling members from various African and Caribbean countries. Some are from European countries, usually Swiss family members. Others are organized along national and ethnic lines. Remarkably, many choristers and choir leaders come from West or Central African regions, specifically from the Congo and Cameroon. As a result, hymns in Beti, Boulu, or Lingala and Kikongo, local languages of Cameroon and the Congo, form the bulk of the musical repertoire. In the final mass, these hymns, which most of the participating choirs now know by heart, build the climax of the pilgrimages. The preponderance of certain ethnomusical repertoires is partly due to the size of an ethnic diaspora, but it also reflects the cultural capital and the transnational links that are mobilized by migrants. The *Association Camerounaise de paroisse St. Joseph de Lausanne* invites musicians like Gaby Ndongo and Luca le Mignon from Cameroon, two artists who perform religious music. In 2012, Gaby Ndongo participated as lead singer in the pilgrimage in St. Maurice. More often, these musicians perform what they call "religious Bikutsi music," a music and dance style of the Beti ethnic group in Cameroon to which many of the association's members belong. In contrast to native Bikutsi, religious Bikutsi dance music in Switzerland is devoid of explicit and sexual allusions. Despite these restrictions, the two musicians appear regularly on stage in a nightclub in downtown Lausanne.[11] The club is frequented by members of the Cameroonian association and diaspora, especially when transnational artists arrive. The same association supports its members in social ceremonies, like "death celebrations" by collecting funds for the transfer of corpses, and other logistical and organizational matters. It is, therefore, less than a typical choir, insofar as its members do not hold regular rehearsals, but draws on a standard musical and cultural repertoire for various events. However, it is more than just a choir, insofar as it mobilizes a variety of transnational networks to consolidate and support the group. In this way, the association works much like urban choirs in Cameroon, travelling to rural places with urban music and performing in religious rituals and social events, thereby preaching the gospel through religious music.

The musical construction of the pilgrims' identities and belonging is performed through a variety of means: costumes, African vernacular languages, instruments like the balafon and the electric organ, and through rhythms and melodies of the pilgrims' musical repertoire. As a social and religious event, the pilgrimages bring together a mixed group of old and new immigrants. They also attract the group of officials, clerics and laity of the Catholic Church who attend to the religious accommodation and incorporation of these migrants in Switzerland. Further, others stakeholders observe and participate in the events. These include: Swiss and other European pilgrims, either missionaries (from *Missio* or from the White Fathers, and the White Sisters, the female counterpart of the Missionaries of Africa), relatives or friends of the African pilgrims, local residents, tourists, anthropologists and journalists.

I now come back to my question of why these events are located in rural places far from Switzerland's larger cities, and why they attract urban choirs to perform "African chorale music"? Through pilgrimage, the Swiss Catholic organs of mission (*Missio*) and migration (*Migratio*) bring together immigrants from Africa, who are otherwise difficult to reach and difficult to identify as Catholics. Most of these immigrants live in Switzerland's larger cities or in the urbanesque surroundings. Few, as I was told by Marco Schmid, a representative of *Migratio*, are willing to be registered. In order to incorporate these immigrants into the

Swiss Catholic Church, symbols and places for identification had to be invented. This was achieved through the construction of religious events centering on religious identity figures, such as the Africans martyrs and saints in St. Maurice, and the Virgin Mary in Einsiedeln. These religious events are symbolically powerful in building ties between the Catholic migration organ, the missionary organizations, and the independent abbeys of St. Maurice and Einsiedeln. The latter have a privileged position in the Swiss Catholic Church, each counting as a diocese headed by an archbishop. The abbeys are therefore powerful institutions in the Catholic Church structure. They are rural sacred sites with considerable urban influence. Incorporating "African choirs," and their exotic performances into the annual calendar of pilgrimages, draws an increasingly significant part of the Swiss immigrant population into the Catholic Church. It also makes them visible in public. In addition, the event is considered by the missionaries and clergy in St. Maurice as being a "reverse mission" by African Christians, who it is hoped inspire Europeans with their spirituality and religiosity. Although this cliché of a potent and lively "African spirituality" (Salzbrunn and von Weichs 2013; Ballif 2010, 2014) is upheld by both sides, missionaries and clergy of non-African origin recall the ancient Catholic warning against "artistic excess in worship music" (Marini 2012). When "African choirs" in St. Maurice and Einsiedeln enjoy singing and dancing in liturgical mass, they are on the one hand cherished for their effervescence and *esprit*. On the other hand, these performances are suspected to cause a loss of bodily control. They thus recall the Augustinian fear of distraction and seduction through sacred music. Such warnings are closely connected to admonitions by the Abbot of St. Maurice against the worship of other gods and spirits, a suspicion that is not entirely implausible, since the idea of exclusive belonging to the Catholic Church is not shared by all pilgrims. Some visit other churches, while others consult religious practitioners like spiritual healers.[12] By traveling to the rural sacred sites of the abbeys and performing religious music, the "African pilgrims" transgress boundaries between the urban and the rural. These boundaries might not exist in the same way in their homelands, as I have argued in the case of Cameroon. At the same time, an "African space" that is both imaginary and real is created and invented by missionaries, clergy and pilgrims.

Conclusion

Investigating the formative and transformative aspects of music and performance in urban and rural religious places by taking a transnational perspective on "religious music" in Cameroon and "African pilgrimages" in Switzerland has led to several insights. I argue with Stokes that music is not only an identity marker in migration processes, but also a means to transgress boundaries and to connect old and new identities, cultural practices, and places. This is the case when rural migrants in Cameroon arrive in town and join one of the many cultural groups and choirs, or when urban residents affiliate with a church through membership in a choir. The new popularity of choral and "religious" music has spiritualized the city and created a market for religious music. Vice versa, urban religiosity is brought to the villages through urban choir music to evangelize and to perform in festive events and rituals. Moreover, the mediatization of "religious music" and the appropriation of choirs and composers for political agendas has helped to legitimize the political regime in Cameroon. New technologies and the liberalization of (religious) associations and assemblies since the 1990s have created new forms of urban religious practices and evangelization in Cameroon with choir music as a medium of the gospel.

Likewise, the recent formation of religious choirs by migrants from Africa is a new phenomenon in Swiss cities. Similar to the growing popularity of African American gospel choirs in Europe, these choirs are exotic and are welcome attractions in church services and congregational events. African immigration in Switzerland is, first and foremost, an urban phenomenon, although migrants also settled in rural and semi-urban areas, when they intermarried or when they lived in homes for refugees, for example. For many of the new migrants, African urban choirs facilitate their incorporation into local Church structures and migrant associations, especially when formal structures of incorporation are lacking or not considered as trustworthy. In Cameroon and Switzerland, old and new homes are linked through urban choir music and cultural performances in urban, semi-urban and rural places. Religious events such as the "African Pilgrimages" in Switzerland were invented by the Swiss Catholic Church to incorporate African migrants, many of whom live in urban and urbanesque habitats. These events innovate the long-established and dominant belief system of the Catholic Church through new practices of "prayer." In these practices, singing is considered as "prayer two times." Further, spiritual sensation is expressed through dance and body movement. Likewise, music and singing is used to re-evangelize white Christians in Switzerland and thus to "reverse mission." However, European missionaries and clergy continue to contain the emotional and bodily sensations of music that threaten the Catholic understanding of sacred music and worship. The different interpretations of "singing as prayer" create a desire for and tension between old and new cultural and religious practices, especially in light of a growing diversity of religious belonging and of religiosity.

The post-colonial city, an urban space marked by rapid urbanization and globalization, functions as a laboratory for religious movements, new religious practices, and musical creativity. Music, especially urban "religious music" in Cameroon has brought new dynamics into modern city life and its urban religious practices. Its incorporative and transformative power is particularly tangible in situations of translocal and transnational migration, especially when place-making is difficult to achieve and migrants live dispersed in cities and urbanesque habitats. In such processes of localization, the semi-rural (ancient Catholic abbeys) sites of Swiss African pilgrimages turn into a stage for (in the Swiss cultural context) new musical and religious practices, such as the performance of Cameroonian choir music.

Notes

1 I would like to thank Monika Salzbrunn, Janine Dahinden, Josephus Platenkamp, Antoine Socpa, Délphine Nanga Ondoua, François Akono Ondoua and Rémy Netongo Ndolo, Edmée Ballif, Alexandre Grandjean and Cécile Navarro for their support and discussions in this study. Likewise my thanks go to the reviewers of this chapter for their valuable comments.

2 See, for example, http://www.augnet.org/default.asp?ipageid=428; accessed August 31, 2015), or http://amazingcatechists.com/2010/07/praying-twice/; accessed August 31, 2015). Ostrem and Peterson point to the selective reception of Augustine's writings on music up to the present, "Augustinus is frequently included in enumerations of church authorities that have had qualms about music but in the end approved it. The statement that Augustinus 'approves of singing in church, so that by the delights of the ear the weaker minds may be stimulated to a devotional mood' (confessions 10.33.50), is quoted again and again . . . " (2013: 1431).

3 This study was part of my twelve-month research project on "Travelling Cosmologies and the Transnationalisation of Religious Worldviews in Europe and Africa" in 2011. This research was financed by the Rectors' Conference of the Swiss Universities (CRUS), the University of Lausanne (UNIL), and the University of Neuchâtel (UNINE) as support for a double career program of the UNIL. I was attached as a research fellow to the *Laboratoire d'*études transnationales et des processus sociaux at the *Maison d'analyse des processus sociaux (MAPS)* of the University of Neuchâtel. I would like to express my sincere gratitude for the support provided by these institutions, and by my colleagues of the MAPS, and of the ISSRC (*Institut de science sociale des religions contemporaines*) of the University of Lausanne.

4 Professor Mendo Zé retired in 2005 as managing director of CRTV. He was protected by the presidential family until 2014, when he was arrested on suspicion of embezzlement of public resources. Journalists suspect that he was sacrificed in the wake of the political fall of Blaise Compaoré in Burkina Faso, a popular movement that could spill over to Cameroon: http://www.camer.be/37138/30:27/cameroun--arrestation-de-gervais-mendo-ze-paul-biya-abat-ses-dernieres-cartes-cameroon.html; accessed November 29, 2014.

5 Interview with Professor Mendo Zé in his residence at Yaoundé, February 2, 2012.

6 www.peuplesawa.com/fr/bnlogik.php?bnid=51andbnk=14andbnrub=1; accessed December 2, 2014); http://www.camerpost.com/le-kaba-ngondo-pas-vraiment-camerounais; accessed June 10, 2015.

7 Constitution on the Sacred Liturgy Sacrosanctum Concilium, solemnly promulgated by His Holiness Pope Paul VI, on December 4, 1963 (see Chapter VI, paragraph 119 and 120): http://www.vatican.va/archive/hist_councils/ii_vatican_council/documents/vat-ii_const_19631204_sacrosanctum-concilium_en.html#_ftn42; accessed September 11, 2015.

8 Fieldwork was done in the same context as mentioned in note 2. In 2011, I visited for the first time the tenth *Pélèrinage aux Saints d'Afrique* to the Augustine monastery in St. Maurice, canton Valais, and the first *Afrikanische Pilgerfahrt* to the Benedictine monastery in Einsiedeln, canton Schwyz. In the following years, I repeated the pilgrimages with the exception of the pilgrimage in Einsiedeln in 2013 and part of the pilgrimage in St. Maurice in 2015. In these cases, I was replaced by Alexandre Grandjean and Cécile Navarro from the ISSRC. In addition to these field sites, I visited African choir events in Bern and Lausanne. My participant observation was supported by informal talks during and after the events with choir members and organizers.

9 After teaching social and cultural anthropology at the University of Münster (Germany), I came to Switzerland in 2011 (see notes 2 and 6) as a research fellow at the University of Neuchâtel.

10 The *Groupment Coopération Missionnaire en Suissse Romande* (GCMSR) is led by the Conference of Swiss Bishops: http://www.cath.ch/blogsf/cooperation-missionnaire; accessed June 26, 2015.
11 Personal communication with Gaby Ndongo and Luca le Mignon at the *Pèlerinage de saintes et saints d'Afrique*, St. Maurice, July 3, 2012.
12 Personal information by a pilgrim of the *Pèlerinage de saints et saintes d'Afrique*, St. Maurice, July 3, 2012.

References

Adogame, Afe (2013), *The African Christian Diaspora: New Currents and Emerging Trends in World Christianity*. London and New York: Bloomsbury.

Ballif, Edmée (2010), St. Maurice, Afrique: construction des identités collectives autour du pèlerinage aux saintes et saints d'Afrique à St. Maurice, Masterthesis, Lausanne.

Ballif, Edmée (2014), Les Africains, origine et avenir de l'Eglise catholique suisse? Réflexions sur les discours des organisateurs du Pèlerinage aux Saintes et Saints d'Afrique à St-Maurice, *Ethnographiques.org*: 1–16.

Becci, Irene, Marian Burchardt and José Casanova (eds) (2013), *Topographies of Faith: Religion in Urban Spaces*. Boston, MA: Brill.

Berliner, Paul (1975/76), Music and Spirit Possession at a Shona Bira, *African Music*, 5(4): 130–39.

Campbell, Heidi (2010), *When Religion Meets New Media*. London and New York: Routledge.

Coleman, Simon and John Eade (eds) (2004), *Reframing Pilgrimage: Cultures in Motion*. London and New York: Routledge.

Connell, John and Chris Gibson (2003), *Sound Tracks. Popular Music, Identity and Place*. London and New York: Routledge.

De Beer, Rudolf and Wilson Shitandi (2014), Choral Music in Africa, in *The Cambridge Companion to Choral Music Online*. Cambridge: Cambridge University Press.

Falola, Toyin and Tyler Fleming (2012), *Music, Performance and African Identities*. New York: Routledge.

Fischer-Lichte, Erika (2012), *Performativität. Eine Einführung*. Bielefeld: Transcript.

Glick Schiller, Nina (2009), There is no Power Except for God: Locality, Global Christianity and Immigrant Transnational Incorporation, in Bertram Turner and Thomas G. Kirsch (eds), *Permutations of Order. Religion and Law as Contested Souvereignties*. Farnham: Ashgate.

Hermkens, Anna-Karina, Willy Jansen and Catrien Notermans (eds) (2009), *Moved by Mary. The Power of Pilgrimage in the Modern World*. Farnham and Burlington, VT: Ashgate.

Hüwelmeier, Gertrud and Kristine Krause (eds) (2010), *Traveling Spirits: Migrants, Markets and Mobilities*. New York: Routledge.

Jindra, Michael (2011), The Rise of "Death Celebrations" in the Cameroon Grassfields, in Michael Jindra and Joël Noret (eds), *Funerals in Africa. Explorations of a Social Phenomenon*. New York: Berghahn.

Kiwan, Nadia and Ulrike H. Meinhof (2011), *Cultural Globalization through Music*. New York: Palgrave Macmillan.

Krings, Matthias and Onokoome Okome (2013), *Global Nollywood: The Transnational Dimensions of an African Video Film Industry*. Bloomington: Indiana University Press.

Livezey, Lowell W. (2000), *Faith in the City: Public Religion and Urban Transformation*. New York: New York University Press.

Marini, Stephen A. (2012), Sacred Music. *Grove Music Online*. Oxford: Oxford University Press.

Martinello, Marco and Jean-Michel Lafleur (2008), Ethnic Minorities' Cultural Practices as Forms of Political Expression: A Review of the Literature and a Theoretical Discussion on Music, *Journal of Ethnic and Migration Studies*, 34: 1191–215, online publication.

Merriam, Alan (1964), *The Anthropology of Music*. Chicago, IL: Northwestern University Press.

Orsi, Robert A. (1999), *Gods of the City: Religion and the American Urban Landscape*. Bloomington: Indiana University Press.

Ostrem, Eyolf and Nils H. Petersen (2013), Music in Augustine's Thought, in *The Oxford Guide to the Historical Reception of Augustine*. Oxford: Oxford University Press.

Pinxten, Rik, and Lisa Dikomitis (2009), *When God Comes to Town: Religious Traditions in Urban Contexts*. New York: Berghahn.

Rosenwein, Barbara H. (2000), Perennial Prayer at Agaune, in Sharon Farmer and Barbara H. Rosenwein (eds), *Monks and Nuns, Saints and Outcasts*. Ithaca, NY: Cornell University Press.

Rytter, Mikkel and Karen Fog Olwig (2011), *Mobile Bodies, Mobile Souls: Family, Religion and Migration in a Global Works*. Aarhus: Aarhus University Press.

Salzbrunn, Monika (2014), *Vielfalt/Diversität*. Bielefeld: Transcript.

Salzbrunn, Monika and Raphaela von Weichs (2013), Sacred Music, Sacred Journeys: What Makes an Event Postcolonial? *ThéoRèmes*, 4: 1–11.

Shankar, Shobana and Afe Adogame (eds) (2012), *Religion on the Move! New Dynamics of Religious Expression in a Global World.* Boston, MA: Brill.

SPI (Schweizerisches Pastoralsoziologisches Institut) St. Gallen. http://www.spi-stgallen.ch/documents/paper%20kirchenstatistik%20dezember%202014.pdf; accessed April 28, 2015.

Stokes, Martin (ed.) (1994), *Ethnicity, Identity and Music. The Musical Construction of Place*. Oxford: Berg.

Swiss Federal Statistical Office/Schweizerisches Bundesamt für Statistik (2003). Eidgenössische Volkszählung 2000. Bevölkerungsstruktur, Hauptsprache und Religion.

Index